W9-ALJ-264

ROYAL PALMS IN FAHKAHATCHEE STRAND

A FRESH-WATER SAWGRASS MARSH IN THE EVERGLADES

A MANGROVE TREE GROWING ON A FLORIDA BAY SANDBAR

A FIRE RAGING ON A PINELAND RIDGE

EGRETS, HERONS AND OTHER EVERGLADES BIRDS

A JUMBLE OF SEA SHELLS ON A CAPE SABLE BEACH

A NEST OF BABY ANHINGAS IN A MOSS-HUNG CYPRESS

VERDANT KEYS SCATTERED ACROSS FLORIDA BAY

LIFE WORLD LIBRARY
LIFE NATURE LIBRARY
TIME READING PROGRAM
THE LIFE HISTORY OF THE UNITED STATES
LIFE SCIENCE LIBRARY
GREAT AGES OF MAN
TIME-LIFE LIBRARY OF ART
TIME-LIFE LIBRARY OF AMERICA
FOODS OF THE WORLD
THIS FABULOUS CENTURY
LIFE LIBRARY OF PHOTOGRAPHY
THE TIME-LIFE ENCYCLOPEDIA OF GARDENING
THE AMERICAN WILDERNESS
THE EMERGENCE OF MAN
THE OLD WEST
FAMILY LIBRARY:
THE TIME-LIFE BOOK OF FAMILY FINANCE
THE TIME-LIFE FAMILY LEGAL GUIDE

THE EVERGLADES

THE AMERICAN WILDERNESS/TIME-LIFE BOOKS/NEW YORK

BY ARCHIE CARR

AND THE EDITORS OF TIME-LIFE BOOKS

Published simultaneously in Canada.
Library of Congress catalogue card number 72-96548.

Valuable assistance was given by the following departments and individuals of Time Inc.: Editorial Production, Norman Airey, Nicholas Costino Jr.; Library, Peter Draz; Picture Collection, Doris O'Neil; Photographic Laboratory, George Karas; TIME-LIFE News Service, Murray J. Gart; Correspondents Margot Hapgood (London) and Jane Rieker (Miami).

The Author: Archie Carr is a graduate research professor in the zoology department of the University of Florida, where he has taught since 1937. An internationally known biologist, he is the author of nine books, including *The Reptiles* and *The Land and Wildlife of Africa* in the LIFE Nature Library. He has also written numerous scientific papers and several short stories, one of which won the O. Henry Award in 1956. Dr. Carr has also received the John Burroughs medal for exemplary nature writing and the Daniel Giraud Elliott medal for preeminence in zoology.

The Consultant: William B. Robertson Jr. is a research biologist on the staff of Everglades National Park, and a specialist in animal populations. For his Ph.D. degree at the University of Illinois (in 1955), he wrote his doctoral dissertation on the land birds of southern Florida. After a stint on the Illinois Natural History Survey, he returned to Florida in 1956. One of his current interests is a study of the endangered bald-eagle population of the park.

The Cover: A southern Florida creek, framed by the fronds of a cabbage palm and bordered by a red mangrove at right, laps the edge of a plain of sawgrass, the ubiquitous ground cover of the Everglades. A clump of cabbage palms partially obscures a "tree island" of hardwoods on the horizon.

Contents

A Low-lying, Watery Wilderness

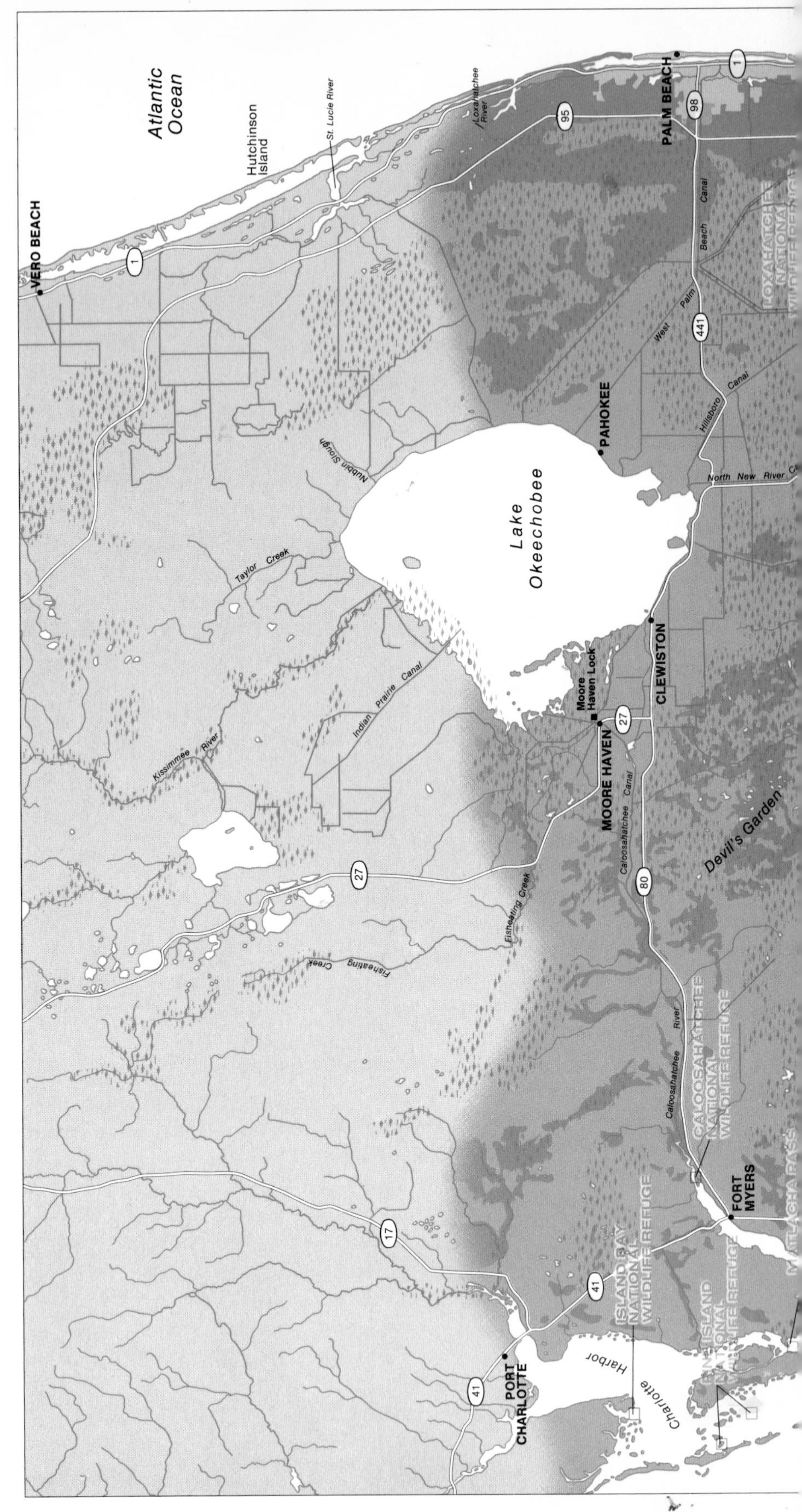

The Everglades region of southern Florida (green rectangle above) includes some 13,000 square miles of wilderness, shown in shades of green in the detailed relief map at right. The predominant terrain is marshy grassland, represented by dotted green areas. Wooded tracts are shown in darker green; urban areas in yellow.

The entire region is low and flat, making for poor drainage. The interior, drenched by 50 to 60 inches of annual rainfall, remains largely under water most of the year. However, the land tilts slightly downward from the region north of Lake Okeechobee (map on page 22); this natural declivity keeps the peninsula draining southward at a rate that has been accelerated by the digging of canals (straight green lines south of Lake Okeechobee and blue lines north of the lake).

Black squares indicate points of special interest; a line of blue dots at bottom left traces the Wilderness Waterway, a 100-mile-long route for small boats. Red lines mark boundaries of parks and wildlife refuges. Black lines mark older roads, including the Janes road (at left) and Loop Road (center) traveled by the author.

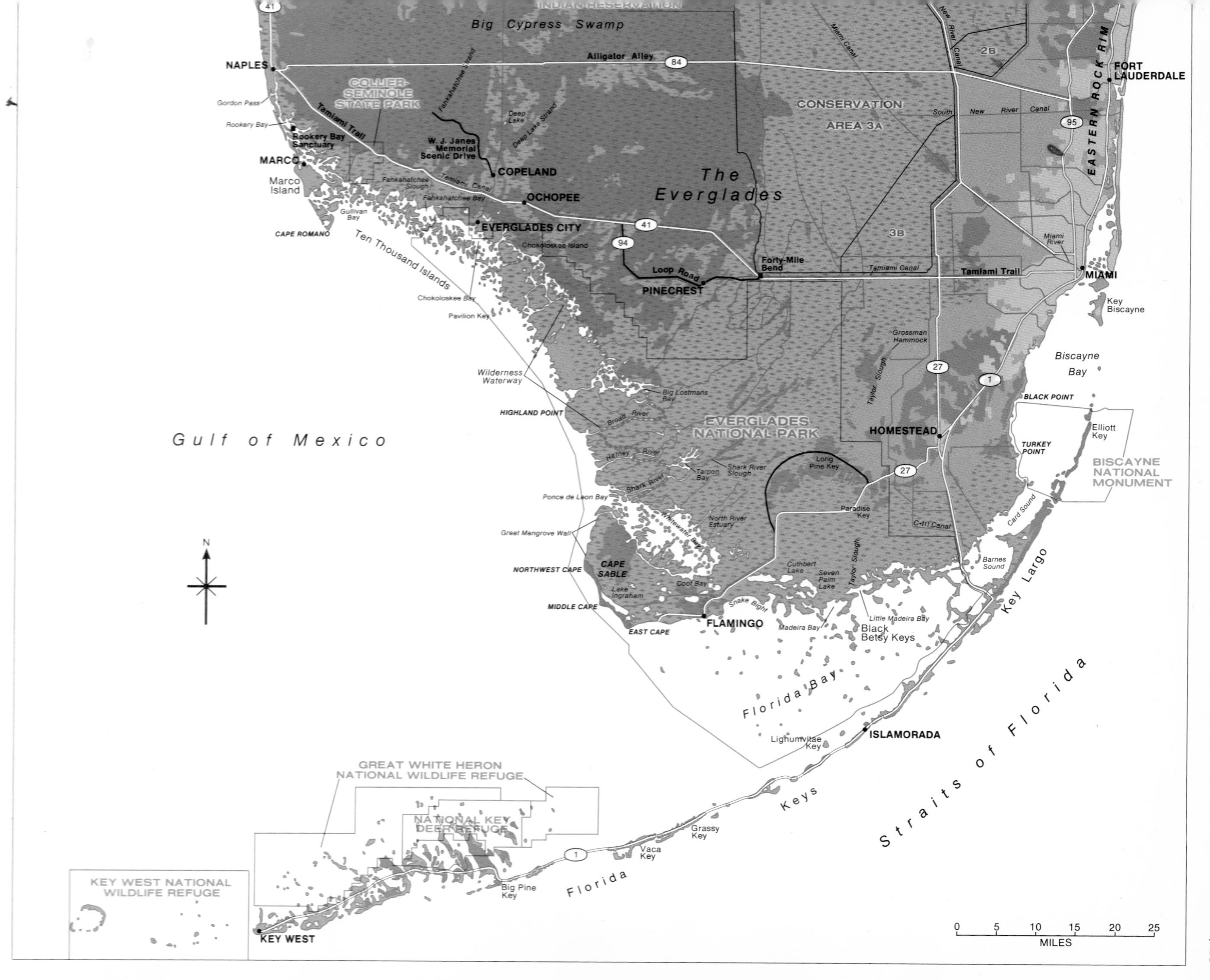
Big Cypress Swamp
NAPLES
Alligator Alley
84
41
Gordon Pass
Rookery Bay
Rookery Bay Sanctuary
COLLIER-SEMINOLE STATE PARK
Tamiami Trail
MARCO
Marco Island
Fahkahatchee Strand
Deep Lake
Deep Lake Strand
W. J. Janes Memorial Scenic Drive
COPELAND
Fahkahatchee Slough
Tamiami Canal
Fahkahatchee Bay
OCHOPEE
Gullivan Bay
CAPE ROMANO
EVERGLADES CITY
Chokoloskee Island
Ten Thousand Islands
Chokoloskee Bay
Pavilion Key
The Everglades
94
Loop Road
PINECREST
Forty-Mile Bend
CONSERVATION AREA 3A
3B
2B
South
New River Canal
Miami Canal
95
EASTERN ROCK RIM
FORT LAUDERDALE
Miami River
Tamiami Canal
Tamiami Trail
MIAMI
Key Biscayne
Biscayne Bay
Grossman Hammock
Taylor Slough
27
1
BLACK POINT
TURKEY POINT
Elliott Key
BISCAYNE NATIONAL MONUMENT
HOMESTEAD
Wilderness Waterway
Big Lostmans Bay
HIGHLAND POINT
Broad River
EVERGLADES NATIONAL PARK
Harney River
Tarpon Bay
Shark River Slough
Long Pine Key
Shark River
Ponce de Leon Bay
Great Mangrove Wall
North River Estuary
Whitewater Bay
Paradise Key
C-111 Canal
Card Sound
Barnes Sound
Key Largo
NORTHWEST CAPE
CAPE SABLE
MIDDLE CAPE
Lake Ingraham
Coot Bay
Cuthbert Lake
Seven Palm Lake
EAST CAPE
FLAMINGO
Snake Bight
Madeira Bay
Little Madeira Bay
Black Betsy Keys
Gulf of Mexico
N
Florida Bay
Lignumvitae Key
ISLAMORADA
Straits of Florida
Keys
GREAT WHITE HERON NATIONAL WILDLIFE REFUGE
NATIONAL KEY DEER REFUGE
Grassy Key
Vaca Key
Big Pine Key
Florida
KEY WEST NATIONAL WILDLIFE REFUGE
KEY WEST
0
5
10
15
20
25
MILES

1/ The River and the Plain

Here are no lofty peaks seeking the sky, no mighty glaciers or rushing streams.... Here is land, tranquil in its quiet beauty, serving not as the source of water but as the last receiver of it.

HARRY S. TRUMAN/ *ADDRESS AT DEDICATION OF EVERGLADES NATIONAL PARK*

Not long ago my wife and I spent a day cruising a network of roads in central Florida near the city of Orlando. We were looking for the beginnings of the Everglades. It might seem an odd place to be searching for the beginning of a wilderness. Traffic whizzed past; Walt Disney World was nearby; and strictly speaking, the Everglades were away down south of Lake Okeechobee, a hundred miles from where we were.

But ecologically the Everglades are linked with a long series of lakes and streams that begins halfway up the interior of peninsular Florida, on the southernmost slopes of the high mid-section of the state known as the Central Ridge. From there this hydrologic system fans out southward through much of the lower half of the peninsula, extending down the Kissimmee River Valley, into Lake Okeechobee and then out across a vast, gently sloping plain covered with sawgrass (a sedge up to 12 feet tall, with leaves set with fine sharp teeth along the midrib and edges). This enormous spread of grass, punctuated with little clumps of hardwood trees—known as tree islands—extends to the bottom of the peninsula, to the coastal mangrove fringe, the sandy storm dunes, and the brackish estuaries of Florida Bay and the Gulf of Mexico.

The southwesternmost corner of this region is the 2,020-square-mile Everglades National Park, the third largest national park in the country. Though it seems horizon-wide when you look out across it, the area it covers is actually only a corner of the whole Everglades, which spread back northeast to Lake Okeechobee and beyond, for 125 miles. To the north and west of the park lies Big Cypress Swamp, 2,400 square miles of complex landscape so closely linked with the Everglades that in places it is hard to tell where one begins and the other leaves off.

Though the sawgrass plain is the heart of the Everglades and though millions of people have visited the Glades—as Floridians call them—queer notions about the place persist. Perhaps influenced by what they have heard about Big Cypress, visitors still arrive expecting to see a dim, mysterious swamp-forest full of reptiles, exotic birds and eerie noises, a sort of Hollywood jungle set in which boa constrictors or a band of apes would not come as a total surprise. This probably accounts for the let-down look park rangers see in the faces of some newcomers as they gaze out over the sawgrass plain for the first time.

But the Everglades are unique: they have no counterpart anywhere on earth. Although the region is almost perfectly flat, few landscapes anywhere have a more intricate interplay of physical and biological factors. The plain is old sea bottom—or, more accurately, new sea bottom, since the whole sawgrass region was exposed only a few thousand years ago by the last retreat of the sea during an ice-age buildup. No folding or warping of the sea bottom was involved. The water simply drained off as the ice at the North and South Poles built up, and the limestone bottom came out as flat as a table top. The land it made tilts barely enough to keep water flowing across it, when there is any water there and when no man-made obstacles intervene. It is this flow that integrates the southern tip of Florida with Lake Okeechobee and the Kissimmee Valley into one vast ecological entity.

If one has the time, the best way to appreciate this unified structure is to follow the drainage gradient—the excruciatingly slow descent of the land from the southern end of the Central Ridge to the Gulf of Mexico on the west, to Florida Bay on the south and to the Atlantic on the east. I had decided to trace out as much of the drainage system as I could without swimming through any culverts, and that is what my wife and I were doing southwest of Orlando that day I mentioned.

We had spent hours searching, and by late afternoon had moved so near the origins of the system that going farther was really superfluous; but the maps we had with us carried the headwaters farther still, and the gap was a challenge. The drainage of this stretch of country, however, was only vaguely shown, and it is risky to indulge a sudden whim to leave the freeway there to see where some creek comes from.

After three hours of combing the area, dodging traffic and denouncing absent map makers for their discordant views of Orange County geography, we finally fixed on a body of water called Turkey Lake as what we were looking for—as the real beginning of the Everglades. The lake was clearly in the Shingle Creek drainage system that we knew flowed southward into Lake Tohopekaliga, the northernmost of the big lakes in the Kissimmee Valley chain. But even after we decided that Turkey Lake was what we were after, we could find no access to the lake shore from any of the roads around it. None, that is, except by vague two-track trails that faded out in the woods around the shore. Finally, in desperation, I turned off into one of these trails, and we quickly found ourselves blocked by a garbage dump that spread for acres through what had once been a fine live-oak hammock. We parked and went on from there on foot, and after walking a hundred yards or so, came out suddenly at the edge of Turkey Lake.

It was a good lake; you could see that at once. Somehow, it had escaped the fate of the woods around it. The water was clean, fish were striking, and somewhere a kingfisher rattled. I took off my shoes and waded out through a narrow zone of marshy shore to open water, flushing a banded water snake along the way. There were patches of applesnail eggs on the cattail stems. Farther down the shoreline a Louisiana heron rose, and up the other way in a cypress tree a snakebird was preening. Though it was a fair, warm Sunday afternoon, there was only one boat on the lake, with a man in it quietly casting for bass. I wondered what curious quirk in the trend of the times had saved a good lake like that from the usual fate of suburban waters. Finding it there added joy to our pilgrimage to the upper limits of the Everglades, and I hoped this augured well for the future of the region.

Looking down into the water around my feet, I saw a posse of gambusias, six little mosquito fish an inch or so long—pusselguts, they call them in northern Florida. There were two males in the school and four females. One of the males was lording it over the other fish, and I fell to musing that back before Florida got into the hands of developers and engineers an able little fish like that, if charged with supernatural drive, might have made his way from Turkey Lake clear down to wherever salt water stopped him at the end of the Everglades. And even today, I thought, if he followed the man-made changes in the route he still might complete the journey. Anyway, he could do it with much less trouble than I could; and the thought came to me that I ought to let him try. There in Turkey Lake, far up inside Florida, the fish was still

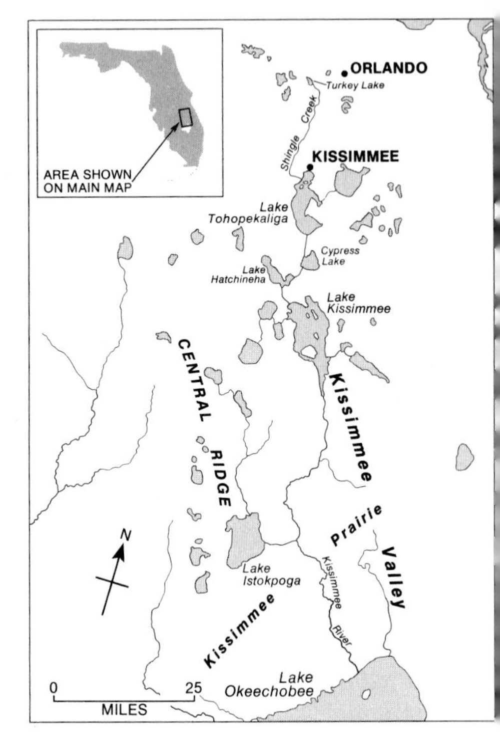

The area shown on this map, which supplements the map on pages 18-19, is a vital part of the ecological system of the Everglades, although it lies to the north of Lake Okeechobee—usually considered the northern boundary of the Glades. The waters that feed the Everglades originate just south of Orlando in the Kissimmee Valley and collect in Lake Okeechobee, before beginning their slow southward journey to the tip of the peninsula.

within the Everglades basin, and in his minuscule way was a working part of its ecology. The slow flow of water bound together all the varied biological communities of southern Florida: the Kissimmee Valley, Lake Okeechobee, Big Cypress Swamp, the sawgrass Glades, the coastal mangrove forest, and even the adjacent waters of Florida Bay and the Gulf of Mexico. All these geographical features were, in a very real way, parts of a single ecologically organized unit.

It was this thought that started me tracing the little gambusia's downstream journey in my mind. I imagined him puttering about the edges of Turkey Lake searching for the outlet, locating it at last, then working his way down ditches, through culverts and across ponds, into Shingle Creek, the first sizable stream of the Everglades basin. From there on, the trail was plain, downstream to the western edge of Lake Tohopekaliga. From here southward there would be some lake hopping to do down the upper Kissimmee Valley. In the days before the way was strewn with levees, locks and dams, fish surely traveled from lake to lake by simply moving when flood times came. Nowadays the route from Tohopekaliga goes through three other big lakes—Cypress, Hatchineha, and Kissimmee—that are connected by canals.

I became intimately acquainted with the route down the chain of lakes one cold, rainy November day a few years ago when I was invited to go on a duck census there. We traveled in an airboat, one of the various gasoline-powered inventions of the devil. For chasing poachers and monitoring duck populations, airboats are useful, I suppose, but they are terrible vehicles all the same—uncomfortable, intrusive and noisy. Anyway, we took an airboat all the way down the chain of lakes and back, on a trip to count the local colony of the mottled duck, a resident mallard and the only duck that nests in peninsular Florida.

The trip began at the city of Kissimmee, on the northern end of Lake Tohopekaliga. We blasted away from the boat ramp at the edge of town and raced down the right-hand shore of the lake, keeping just inside the shoreline vegetation. I noticed that a lot of new flat shore had been exposed since my last visit. For nearly a hundred years the whole Kissimmee Valley has been subjected to drainage operations of one kind or another. The latest, carried out only recently, was a drastic drawdown of Toho's waters, aimed at exposing bottom deposits to oxidation, with the hope of lessening the lake's overenrichment with nutrients from sewage and fertilizers. Part of this pollution comes from the city of Kissimmee; but some also comes from far up the little tributaries of Shingle Creek, even including some near Orlando. The Tohopekaliga water, in turn, pouring southward through the valley, was polluting the lakes farther down the chain and even exacerbating the problems of Lake Okeechobee itself. Thus Orlando dishwater goads Okeechobee algae into intemperate growth—proof that the flow of water integrates the whole Everglades basin into one ecosystem.

All the way down through the chain of lakes, the airboat pilot kept us just inside the fringe of shoreline plants. Where these were rushes or maiden cane we flushed few waterfowl; but wherever the grass gave way to patches of pickerelweed, or to open fields of water lilies, ducks appeared in abundance. Mottled ducks were few and scattered, but there were hundreds of wood ducks, blue-winged teal in thousands, a lot of ringnecks, shovelers and widgeon, and even a few little bunches of greenhead mallards, which one doesn't see often in Florida, or I don't. In some places teal and coot were so thick and so loath to leave that we charged into the splash of their rising. That horizontal rain, added to what came down from the sky, kept us soaking wet. We had lunch in a warm little restaurant beside the lock and dam at the head of the Kissimmee River, then made the whole ranting journey back in a steady rain. The round trip was 150 miles, the mottled duck count was 300. And I had seen for myself that, given long life and a driving wanderlust, a pusselgut could very well travel from the delta of Shingle Creek in western Lake Toho to the head of the Kissimmee River.

A pair of mosquito fish—the two-inch-long female is much larger than the male—scout the water for a meal. Scientifically known as Gambusia, the species was nicknamed for its main food: an adult fish can daily consume its own weight in mosquito larvae.

The Kissimmee River is the chief tributary of Lake Okeechobee and the master stream for a vast complex of landscapes linked to the Everglades system—prairies, marshes, sloughs, hammocks and pine flatwoods. Parts of the Kissimmee Valley look very much like the Everglades—so much so that some of it, the sawgrass and wet prairies in the valley's southern and western sections, is sometimes called the Little Everglades. Before the Kissimmee River was channelized, when it meandered through a hundred miles of meadows, its valley was much wetter than it now is. Nevertheless, early visitors recognized it as promising cattle country, and soon after the Seminole wars—in the 1830s and '40s—the region began to go into ranches. As these spread, artificial drainage increased, and the water table went down markedly. But it is still good country, and parts of it have hardly changed at all.

The flatlands of the Kissimmee basin, including the region loosely known as Kissimmee Prairie, are a part of the coastal lowlands, the series of sandy marine terrace plains that extend down both sides of peninsular Florida and meet northwest of Lake Okeechobee. In the north these terraces are mainly covered with pine flatwoods of several different kinds. In the south they merge with the sawgrass Everglades below Lake Okeechobee and with the tree swamps, glade lands and pine flatwoods of Big Cypress Swamp.

In Florida the word prairie is applied to two different landscapes, neither of which, except in flatness, resembles the dry Western country the term usually brings to mind. In northern Florida where I live, prairies are solution basins—that is, marshy lakes filled with emergent and floating herbaceous plants, and subject to abrupt drying up when water levels drop and the water drains through holes in the limestone bottoms. One season you may fish from a boat in one of these places; the next season you may walk across dry ground there. These northern Florida prairies are similar in general appearance to what are known as wet prairies in the Everglades, where they are one of the landscape variants in the sawgrass plain.

Prairie means something different in the expanse of lowlands above Lake Okeechobee. The prairies here are low, flat, tremendous meadows, set with stands of saw palmettos, copses of live oaks, groves of cabbage palms or mixed hammocks of various temperate and tropical hardwood trees.

The dry prairie is the characteristic country of the Kissimmee Valley region; nowhere on earth is there a terrain just like those short grass and saw-palmetto savannas. A place it brings to mind is the grassy savanna land of eastern Africa. Kissimmee Prairie all but cries out for antelope, but lacking those, it has made do with white-tailed deer, wild Kissimmee ponies and box turtles. It is a land where long-legged birds stalk the uncluttered ground, or walk from pond to pond, and where once I saw wild turkeys, sandhill cranes and wood storks standing together on a burn. The storks were just at the edge of a drying pond, the turkeys and cranes not 50 feet away on fire-blackened grass.

Like most of Florida, the original prairie was almost surely a landscape shaped by fire. As in the sawgrass Everglades—and in most of the big savanna lands of the world—fire is constantly in the background here; although its role is confused by other factors that help to keep the land as grassland. Periodic drought and periodic flooding, for instance, are both involved. Some of the original aspect of Kissimmee Prairie was no doubt due to seasonal soaking and a high water table and some to seasonal drought and baking of the surface. But the most fundamental factor that kept short-grass savanna here was probably the fires that intermittently or regularly crept or swept across the land and prevented the growth of broad-leaved trees except where some special local condition allowed patches of hardwood and palm to develop.

Before the white ranchers came, the prairie was Indian country. The Seminoles loved to burn the land—to flush out game, to kill ticks and rattlesnakes, or to make new grazing land for their cattle and ponies. When white men moved in they did the same, only worse, because they were more uneasy than the Indians about snakes, ticks and the black wolves that howled in the night. But the question is, how prevalent was fire before people of any kind came to the prairie? The answer obviously depends on how often you think lightning hit tinder at times when dry fuel lay on the land. It seems to me that this must have been often, and that you see a sign of it in the avid way each new generation of prairie birds quickly learns to crowd onto newly burned ground to eat the cooked animals there, especially the big grasshoppers that seem too dim-witted to flee from fire.

Two characteristic birds of the prairie community are especially fond of cooked victuals. One of these is Audubon's caracara; another is the Florida sandhill crane. There are two distinct subspecies of this crane in the state. One is migratory; it nests in Michigan and spends the winter in the South. The other is a year-round Florida resident, and its chief remaining stronghold is Kissimmee Prairie. Its feeding habits obviously evolved in a prairie environment, and its predilection for burned ground suggests a long association with fire. Anyway, back in the days when fires were more prevalent than they are today, if you wanted to find cranes the best thing to do was to ride a horse out to where a big smoke had been rising for a day or so.

When you got out to such a burn you nearly always found Audubon's caracara there. The caracara is as intrinsic a part of this landscape as the saw palmetto. Though it doesn't exactly look the part, the caracara is a kind of long-legged hawk. The name caracara is supposed to suggest the creature's voice, a peculiar grating croak that is uttered as the bird tilts its head back between its shoulders.

The third member of the avian triumvirate of Kissimmee Prairie is the burrowing owl. A small bird, standing about nine inches tall, it too has very long legs for its body. While this amiable owl is a characteristic occupant of Kissimmee Prairie, it also turns up in other parts of the peninsula that have a proper prairie look. Much of Florida has recently been put into short-grass savanna, in the form of pastures, golf courses and airports; and these are the habitats the little owls have colonized. Their settling at Miami International Airport stirred the Florida Audubon Society to have the airport designated a wildlife preserve.

Burrowing owls are gregarious, and if you come across one anywhere it is likely to be a member of a colony. The burrows are tunnels that, in Florida at least, usually seem to be scratched out by the owls themselves. They are about three inches high and five wide, and go down at a slant for several feet, the final depth often being determined by the top of the water table. The burrow serves as refuge, dormitory and nesting place. During leisure times outside, the owls stand about on little raised places, and bob up and down engagingly when approached.

Because the prairie landscape in its original form was unusually well suited to human needs, the man-made changes there seem a little less drastic than in other Florida terrain. The whole region is certainly much drier now than it was when the wolves and Indians left it, because of the increased drainage that came with the growth of the human population and the spread of the ranches. At the same time, the frequency of wildfire has been reduced. While this is clearly a blessing for snakes, quail, mice and box turtles, it is bad for cranes and caracaras, depriving them of the grasshopper bakes that brightened the days of their fathers. Both of these birds are as stimulated by the disaster of fire as storks are by the disaster of fish-stranding drought. It may be because of this change that cranes and caracaras are less numerous now in Kissimmee Prairie than, say, back in the 1930s when burning was practically a ceremonial practice among tenant farmers.

But I ought to return to tracking the little mosquito fish in his travels southward. We left him in Lake Tohopekaliga. From there he would continue his trip via the Kissimmee River, which flows out of Toho's southern edge. The descent of the river, now that the canal has sliced across all the bends and meanderings, is a great deal shorter than it used to be. A fish would find it all steady swimming today, with the flowing stream cut into placid segments by six locks. I wonder whether the gambusia swims fast enough, once inside the lock, to leave it in company with the same boat he had to await to enter it. But no matter; just imagine him cruising on down through the cut-up river into 65E, the last Kissimmee lock, and out into Lake Okeechobee, the "Big Water."

Okeechobee is 750 square miles of water—roughly 35 miles long and 30 miles wide, and next to Lake Michigan the biggest fresh-water lake wholly within the contiguous United States. It has an average depth of only 14 or 15 feet. Its most important tributary, after the Kissimmee

River, is Fisheating Creek to the northwest, the name of which is a translation of the Seminole word *thlothlopopkahatchee*—changed, presumably, because it was hard to pronounce.

The lake has no natural outlet. Originally it just flooded out over its southern rim and spread into the Everglades. The southern shore used to be densely forested with a magnificent swamp of buttressed custard-apple trees standing on a deep deposit of peat. So in the old days an itinerant fish could have crossed the lake to its southern shore, waited for the wet season, and then wended its way out through any of a thousand little waterways into the custard-apple swamp. At that time these waterways served as small floodgates, holding back high water in the lake, rationing out the excess, and thus prolonging the productive wet stage of the Everglades' annual cycle of drought and flood.

Our latter-day gambusia, however, would find no downstream way out of Okeechobee anywhere except through man-made structures. After two disastrous hurricanes, one in 1926 and the other in 1928, dikes were built to protect the rich farmlands south of the lake. Water now leaves only under human control: westward to the Gulf of Mexico by way of the Caloosahatchee River, eastward or southward through canals to the Atlantic or into vast artificially impounded reservoirs—water Conservation Area Nos. 1, 2 and 3. These occupy most of the upper two thirds of the old sawgrass Everglades that has not yet been drained and converted into farmland.

The fish could leave Okeechobee at its southeastern edge by way of the Hillsboro Canal, enter Conservation Area No. 1, and from there proceed south through Area Nos. 2 and 3. But a route closer to the primeval drainage pattern would be from the southernmost shore of the lake into the Miami Canal, across the muck farms to Levee No. 5, then out through a culvert into Conservation Area No. 3. This impoundment adjoins the Shark River Slough and Everglades National Park, and delivers to the park whatever ration of overland flow it gets.

The gambusia is now nearing the heart of the Everglades—Pahayokee, as the Seminoles knew it, or the River of Grass, as Marjory Stoneman Douglas called it in her book of that name. The sawgrass spreads over a mat of peat resting on a limestone plain elevated no more than 15 feet above the sea in which it formed. In places the surface of the stone floor is fantastically pitted and craggy, but you see this only when the covering of peat has burned off or been blown away. This is a country so flat that no streams form on it. When heavy local showers come, the

Nearly four feet tall, a pair of wary sandhill cranes—the male's head is the brighter red—survey their nesting area.

rain water collects in low humps and trickles outward through the sawgrass that covers the plain. The dip of the land is so gentle that you almost have to accept on faith that there is any flow at all. No matter how hard you look down into the clear, warm water you see no movement in it, or see only the tiny surge from the waving sawgrass stems. So there is no real current to speed the gambusia as he threads southward through forests of sawgrass, and more open stands of pickerelweed, skirting the scattered tree islands and pausing to consort with others of his kind under lily pads in alligator holes.

Despite the distance he has come, there are few things out here in the sawgrass to surprise or offend a gambusia. It is bigger country than he knew back at Turkey Lake, and warmer, and more densely grown over with grass; but the familiar array of food is reassuring—the teeming tiny crustaceans and insect larvae that keep him constantly fed along the way. It is the talent of gambusias to take the world as they find it anyway. They inhabit big lakes, little ponds, brackish ditches or the water-filled tops of old cypress stumps. They eat any nourishing object they can swallow or break apart, and they bear living young that are able at birth to fend for themselves. And they reinforce this survival pattern with an extraordinary ability to spread about the country. It seems wholly impossible to account for their presence in some of the places where you find them. In fact, gambusia is one of the fish that cause people to speak of fish raining down, or of their traveling in mud on the feet of birds. It has never been proved that gambusias do either, but their ubiquitousness tempts one to think of such exciting ways for fish to travel. In any case, this mosquito fish whose journey we have traced all the way from Orange County into the sawgrass Everglades has faced no serious hardships anywhere along the way. And now, out in the River of Grass, he is in a classic gambusia landscape —or will be when he moves on down to the lower end of Conservation Area No. 3, goes out through one of the spillways that run under the Tamiami Trail, and swims through the willows and cattails into the sawgrass of Shark River Slough and Everglades National Park.

Out there he will have come a long way from Turkey Lake—200 miles or more, as the crow flies—but nowhere along the way has he been in alien surroundings. A 30-pound channel catfish in Lake Okeechobee may have surprised him, to be sure, but mostly he took with aplomb the new faces he met. Had he been a real fish, instead of an imaginary one, he would long before have stopped in some sedge-rimmed pool or bladderwort patch where his kind of hunting was good—where the little crustaceans called *Cyclops* danced like gnats in the amber water; where young gatorfleas frolicked among hanging duckweed roots; where tender beetle larvae slithered along the rush stems, and pink worms cowered in the silt. All down the line he would have been kept alert by the ever-present chance that he himself might be eaten, but always the danger would have been a known one—a warmouth bass charging through the sheltering weeds of water milfoil, a tiger beetle lurching at him, a dragonfly larva groping hopefully up from the bottom, a diving-bell spider plunging down from his hunting stand on a stem of pickerelweed.

Such threats are only the way of the world to a gambusia. So our traveler could have stopped off anywhere along the way and no doubt have lived happily there; but because he exists only in my mind's eye anyway, I will move him on down toward the zone where the sawgrass begins to feel the nearness of the sea, and the creeping flow of water begins to ravel out into a maze of mangrove-bordered creeks and estuaries. When finally the little fish stops, it is in a place 85 miles downstream from Lake Okeechobee. Out in the northwest lies Big Cypress Swamp. Tarpon Bay is due west, and from there the Shark River flows on to the Gulf of Mexico. To the south, through mangroves and hammocks, are the coastal prairie and the shores of Florida Bay.

There, in the last alligator hole that never turns too salty for his taste, where the first few mangrove trees begin to mingle with the thinning sawgrass and the last of the cypresses, the itinerant gambusia falls in with a band of his fellows and stops. The pool in which he ends his voyage was derived from different sources. Part of it was local rain; part came out of Big Cypress Swamp. The tinge of salt in the pool washed up from downstream during storm times. And another part of the water came in from the farthest reaches of the River of Grass, and all those distant places beyond, where the fish, too, began his journey.

Portraits of the Land

PHOTOGRAPHS BY RUSSELL MUNSON

Observed casually from a plane, the 4.3 million acres of southern Florida that includes the Everglades appears to be an endless green and brown sprawl of flat terrain. The only perceptible movements in this vast, generally watery tract are the ripple of trees and grasses in the wind, the shifting of cloud shadows over the earth and the glint of sun on water.

But an attentive eye can detect the subtle clues that identify differences among the region's landscapes. The aerial photographs on the following pages, taken from a low-flying plane at the start of the summer rainy season, comprise a gallery of portraits of the land that bring out these differences separately and precisely.

The five distinct but dovetailing regions shown, and located by dots on the maps accompanying the photographs, are typical of the watery parts of the Everglades and their environs. Starting with the area below Lake Okeechobee, they stretch 100 miles south to Florida Bay. Southwest of the lake lies Big Cypress Swamp, where green domes of pond cypress dominate paler ranks of dwarf cypress *(right)*. The water that is visible from the air hints at the swamp's soggy nature; it is flooded most of the year.

Adjoining Big Cypress is the region of the Shark River Slough *(pages 34-35)*. Though from above it seems a continuation of the swamp, the land is more open. The green bulges are not cypress domes but irregularly contoured islands of relatively fragile coco plums and other hardwood trees. The tan carpet surrounding them is sawgrass, the sturdy sedge marking the true Everglades; and the sawgrass is underlaid and nurtured by a barely moving sheet of shallow fresh water.

In the third region, at the headwaters of the Shark River *(pages 36-37)*, the land from a height looks like an arid steppe. In reality, it is a lush concentration of sawgrass, cut by creeks that have begun to channel the fresh waters from the Shark River Slough on their seaward journey.

Sawgrass plains yield to mangrove jungles and fresh water to salt water in the last two landscapes shown. The Shark River estuary and the Ten Thousand Islands *(pages 38-39)* are areas of ceaseless give-and-take between sea and land, each trying to withstand the encroachment of the other. In the region of Florida Bay *(pages 40-41)*, with its mangrove-fringed keys, or islands, the open sea prevails over the land at last.

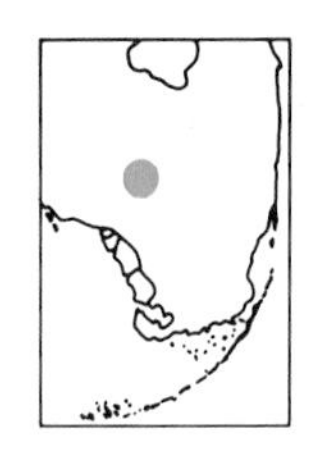

Like verdant craters, domes of pond cypress thrust up from Big Cypress Swamp, surrounded by dwarf cypress. The marl soil of the swamp is often less than a foot above bedrock. But in some places the porous limestone base has crumbled, leaving shallow depressions with rich, soggy soil in which cypresses can root and grow as high as 100 feet —more than 20 times that of their stunted neighbors. The "ponds" in the center of the cypress domes are caused by deeper sinkholes in the limestone.

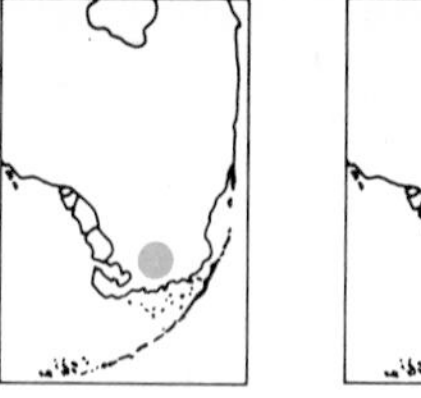

Dense hammocks of tropical hardwood trees punctuate a plain of sawgrass and spike rush on the south side of Shark River Slough. The fresh water of the slough collects in this basin and slows almost to a standstill, enabling seedlings that gained footholds here to evolve into well-defined, roughly circular islands.

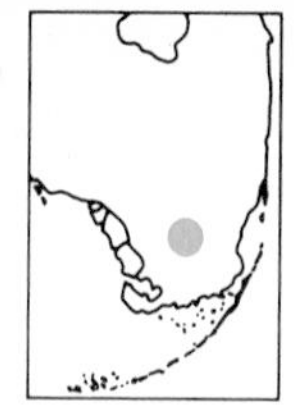

An island of coco-plum trees stands in the mainstream of the Shark River Slough, where the waters—unlike those at the slough's periphery—move at a slow but sure rate. Their erosive power is clearly revealed in the teardrop shape of the island, tapered to a point at its downstream end.

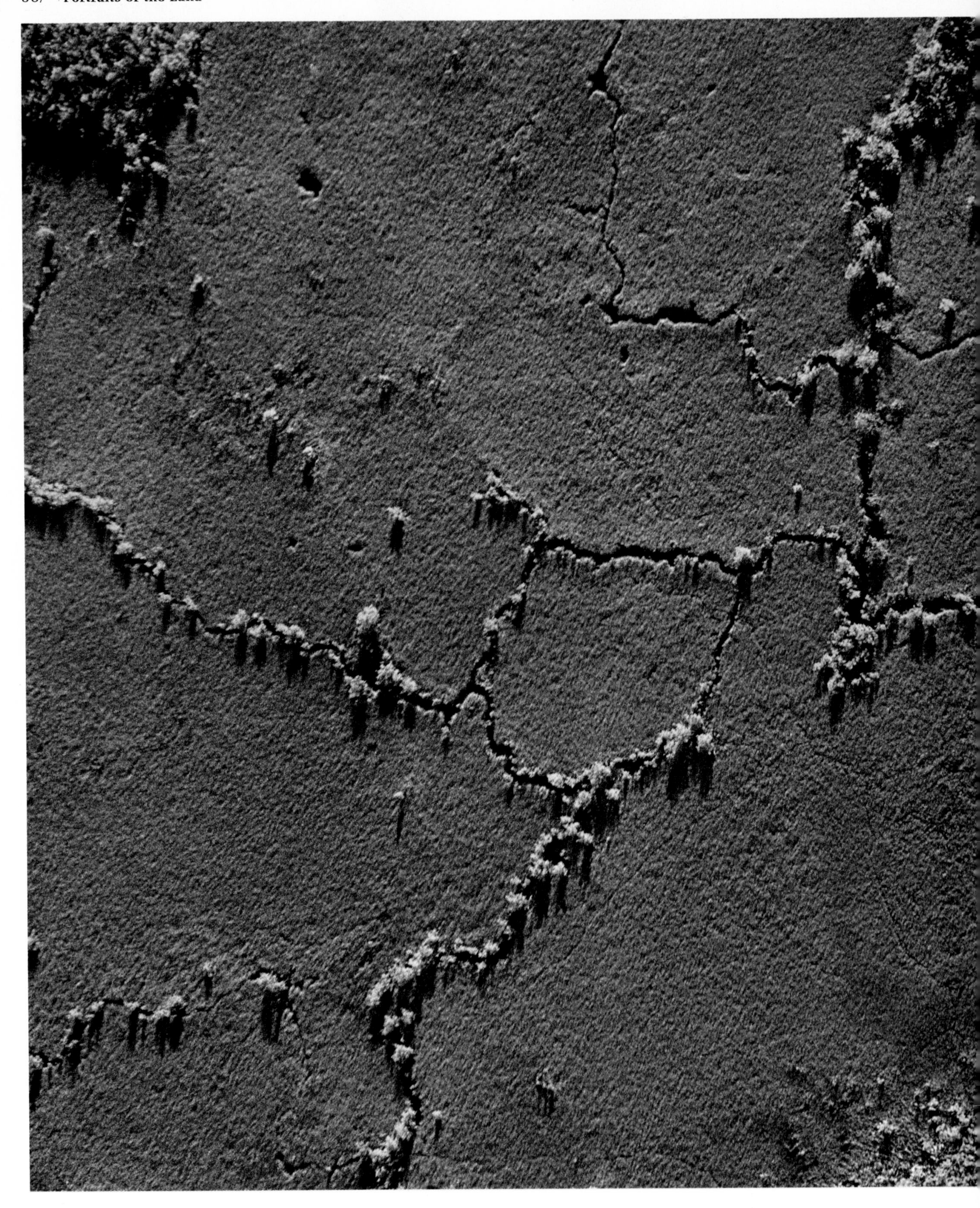

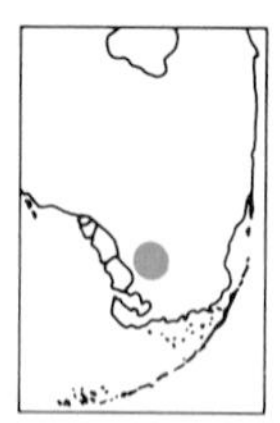

A field of 12-foot-tall sawgrass, so tough it takes a hurricane to lay it flat, flourishes at the headwaters of the Shark River. What appear to be fissures are creeks, issuing from the slough to the northeast; fringing them are mangroves that have invaded from the south. For aquatic creatures to survive, the creeks must be cleared of the mangroves' clogging roots. Here, the prime flow engineer is the alligator, which uses its snout to dredge the channels, tearing away the mangrove roots with its teeth.

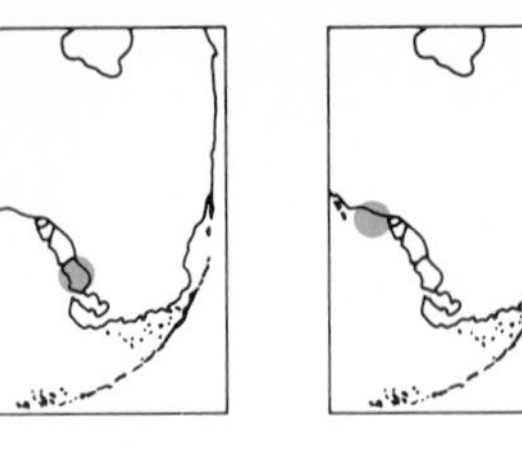

Sunlit waters in the estuary of the Shark River meander in a mangrove maze as they mix with sea water on their way to the nearby Gulf of Mexico. The mangroves here—some as high as 80 feet—are able to resist the inroads of the sea on outposts of land their roots have gathered and continue to hold together.

More vulnerable to violent storms, the Ten Thousand Islands off the Florida coast in the shallows are actually elongated shell bars, some only four feet wide. But despite the sea's recurrent attempt to reclaim these spits of land, their mangrove forests tenaciously hold—and even increase—their acreage.

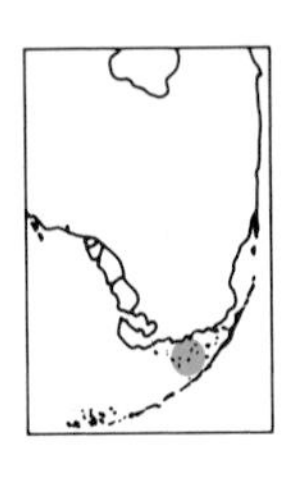

Remnants of a drier epoch, these keys in the shallow waters of Florida Bay mark the last outposts of the Everglades lands. Once hammocks set in a prehistoric swamp, they have struggled against the rising sea over the past 5,000 years, protected against total destruction by the mangroves on their perimeters. The trees continually collect soil and build up a barricade secure against all but the most severe hurricane tides.

2/ The Big Cypress Swamp

Deer... led us through a dream world of gray cypresses and silent Spanish moss and soft knee-deep watery sloughs.... I stood and stared and could not believe that I held orchids in my hands. MARJORIE KINNAN RAWLINGS/ *CROSS CREEK*

It is a habit of mine to file away, more or less retrievably in my memory, recollections of small bits of landscape I have seen—not just big, grand things like mountain views but perfect little fragments of country that I can think back to with a surge of the pleasure the first sight of them brought. One of these places lies within Big Cypress Swamp in southwestern Florida. It is a particular bayhead that I recall—a low evergreen swamp made up chiefly of bay, pop-ash and custard-apple trees—and though it covered only a minute part of the million and a half acres of Big Cypress, it gave me my first real taste of the magic that the entire region holds.

The memory of this bayhead comes down from the time when the Tamiami Trail was new and still sparsely traveled, and a drive across the edge of Big Cypress on a June day was constant excitement. Turtles and young alligators basked along the banks of the canal that parallels the road, and a half acre of herons, egrets and ibises might be spread over any little prairie that you passed. There were lots of snakes along the road shoulders in those days, and at the mouths of the little rills that frothed into the canals after a shower rowdy gangs of short-nosed garfish and soft-shelled turtles crowded, thrashing and squabbling over flotsam that washed in from the Everglades.

This particular bayhead hovered over a dark little stream that crossed the trail somewhere between the towns of Ochopee and Marco. I stopped my car and got out because a speckled six-foot king snake was lying in the road. After communing with him awhile I goaded him into gliding off into the canal and swimming across it into a patch of bushes. I then turned idly to peer down into the sparkling water that was sucking into the culvert at its upstream end. It was clear black like Costa Rican coffee, and I could make out the dim front half of a big snook and a lot of garfish stacked like cordwood, all lurking together in the shadows. I had a casting rod in the car, but I knew that an old, cynical roadside snook would put no stock in my little shiny spoon, or in any other bait short of maybe a big live golden shiner. So I dismissed all thoughts of catching the fish in the culvert. I looked upstream to where the water flowed out of a tunnel in the fringe of moonvines at the edge of the bayhead. Through the tunnel I could see open space back inside under the overhanging trees, and though the water in there was shaded, here and there its surface glistened with quick flashes of gold light. I knew the look of baby tarpon rolling in black swamp water, and I went back to the car and got my rod.

The stream was the only way into the place, so I waded out into it waist deep and walked up through the low opening in the vine tangle. Under the arched ceiling of ash, maple and pond-cypress trees only little rays of sunlight came in, and gleamed green gold and amber on the leaves and glossy water. I could see that if I followed the channel upstream I would pass from one such vaulted chamber to another, all connected by tunnels through the tangled vines over the deeper runs. The little tarpon stopped their frolic as I entered. My rod was useless in there; it was too crowded to cast a spoon. But I didn't care much, because the place was cool and dreamlike and different from any place I had ever been. And besides its look and feel, the scent of ghost orchids in bloom filled the water-floored room like a greenhouse at Easter time. The ghost orchid is a leafless species, with a big waxy white flower that looks, as the botanist Donovan Correll has said, "like a snow-white frog suspended in midair." The flowers are borne on short spikes rising directly from speckled roots that snake about on the trunks of trees, especially pop-ash trees.

There were other things that I recall: a green tree frog sleeping on a vanilla-orchid leaf; a lot of slender striped anole lizards stalking and bobbing on moonvine stems; and a thin rough green snake that rested weightless, motionless and elegant across the tips of a wax-myrtle bough. All these, too, keep that place clear in my memory.

The last time I stopped by that little bayhead a bulldozer had gone through it, and the creek ran milky with silt under a confusion of cat briers. Still later—after droughts and more draining—the country around Ochopee and Marco dried out badly, and a fire roared through the bayhead, and a small but singular and important part of the world had gone. So I cling to that first June day to think back to.

Actually there are a great many such places left in Big Cypress country, even in these lean days. Maybe none with young tarpon flashing silver gold in amber water under ghost-orchid flowers. Perhaps that special juxtaposition required some primeval water pattern that now is gone for good. But if you don't mind wading, there are other vignettes to be stored away in your memory: a troop of young otters roistering in the water under a hanging garden, or a snowy egret standing incandescent in the midday twilight by a black swamp pool in air so still there is never a ripple in the thin lace of its plumes. A person would have to be very shut-minded or badly afraid of snakes not to enjoy wading navel deep through a Big Cypress bayhead. The dim light and moist cool air, the clear black water, the profusion of orchids and other plants growing perched in the trees cast a spell that is at once bizarre and tranquil. And mainly because of the diversity of these plants, no two bayheads are ever just alike.

As a whole, Big Cypress Swamp presents an immensely varied landscape spreading west of the Okeechobee-Everglades basin between the sawgrass plain and the slightly elevated sand ridges of the Gulf Coast. Despite its name, which refers not to the size of its trees but to its great expanse, it is by no means all tree swamp. Its 2,400 square miles include many strikingly different kinds of country: wet prairies and marshes of spike rush or sawgrass; broad savannas set picturesquely with clumps of cabbage palm; and, on the low outcrops of limestone, islands of pine or hammocks of mixed tropical and continental trees. The tree swamps themselves may contain various combinations of trees —ash, custard apple, red maple, willow or oak in addition to the bald cypress and pond cypress, and even these two related kinds of trees make very different-looking kinds of country.

The bald cypress—a deciduous conifer that gets its name from its bare look in winter, after its leaves are shed—thrives best in places with the more lasting water supplies, and with a soil of muck, clay or fine sand. In Big Cypress such places are mainly the sloughs. In local parlance, the word slough is applied to a shallow trough in the limestone floor of the region where drainage concentrates and forms a relatively permanent body of water. The mature bald-cypress tree usually rises out of a buttress near its base, a swelling that may be eight to 10 feet in diameter. This helps the shallow root system support the tree in unstable ground. The roots develop conical offshoots, called knees, that stick up out of the water; these were once thought to serve as respiratory organs, and this is still believed by some botanists, though apparently nobody really understands the function of these knees.

A slow grower, the bald cypress can reach enormous age—some as many as 600 years old have been found—and regularly attains heights of over a hundred feet. The bands of cypress timber that grow in the sloughs are, as in much of the southeastern United States, known as strands. Since the region was first settled in the 1870s, the strands were hungrily contemplated by lumbermen. Logging in a swamp is understandably troublesome, and it was only the great cost and difficulty of timbering in Big Cypress that postponed the ruin of the strands for so long. But by the time of the building boom of the late 1940s the price of cypress lumber, a long-lived wood widely used for roofs, boats, coffins, stadium seats and pickle barrels, had risen spectacularly; the demand warranted the enormous logistic effort and expense required to get it out of the swamps. It has been estimated that 36,000 trainloads of cypress logs were taken out of the Big Cypress area in the 1940s and early 1950s. Many of the trees that were cut down were over a hundred feet high, up to eight feet in diameter above the buttress, and 400, 500 or 600 years old. Practically all of the big bald-cypress timber of Big Cypress Swamp was cut.

Another variety of cypress that has fared better is the pond cypress. Smaller and more widespread than the bald cypress, pond cypress may crowd in tight stands, or scatter over open savannas. It forms the original marginal vegetation of many ponds and streams, and its stands sometimes mix with slash pine.

One striking contribution of pond cypress to the landscape is the cypress dome. This is a roughly circular, dense stand of trees that forms a dome-shaped canopy against the skyline, with the trees tall in the center and evenly decreasing in height toward the edges. It makes a remarkably symmetrical and convex silhouette, and a lot of the domes scattered around, separated by prairie or cleared pineland, give the impression of artifacts of some kind—as if somebody had planted them there with some esoteric sylvicultural plan in mind.

Among the most bizarre landscapes of southern Florida are the dwarf-cypress savannas, tracts of drastically stunted pond-cypress trees growing with sawgrass at the brackish edges of the Everglades. Here the cypresses may be no more than three or four feet tall; afflicted by fluctuating water levels and poor soil, they seem to be growing under almost intolerable conditions. What the dwarf cypresses look like, really, is bonsai trees, growing under the care of some clever Japanese family that every few years comes by and does occult things that keep the trees tiny though big-buttressed and twisted with age. Some people, when they view these little trees, make the mistake of thinking they are cypress saplings, but the spread of their bases tells the true story. In most cases they are much older than the big, ebullient trees that grow tall and dense on better cypress land.

Butterfly orchids, so called because the delicate petals at the center resemble butterfly wings, seem suspended in the forest air—and they almost are. As epiphytes, or air plants, they attach their roots to pond cypress and other trees for support, but they take their nourishment entirely from air, moisture and sunlight.

Pond cypresses are far more favored than bald cypresses as a host for the epiphytes, plants that grow perched on the limb of a tree or clinging to its trunk. The ghost orchids I mentioned in my favorite bayhead are one kind of epiphyte, as are many other species of orchid. Many of them produce colorful, fragrant blossoms. Contrary to popular belief, all air plants are not parasites; most kinds get only support and a place in the sun from the host tree. One kind, however, the strangler fig, is a relentless killer of even very large trees. Often it starts with a tiny seed dropped into a tree by a passing bird. As it sprouts, the little plant fastens itself to the tree and then sends down long roots that take hold in the ground and wrap tightly around the host tree as they gradually strengthen and multiply. Eventually the host is crushed to death, while the strangler fig grows in its place.

The dominant epiphytes in the bayheads and swamps of Big Cypress are types of air plants called bromeliads. As you drive through the region, one of the eye-catching landscapes you will see from the road is a stand of pond cypresses with their trunks bristling with stiff-leaved air plants, which are spectacular when in bloom. If you travel out from Naples toward Fort Lauderdale on Highway 84, called Alligator Alley, you may see the car ahead screech to a perilous stop next to one of these cypress-and-air-plant communities, and somebody in the car will more than likely yell, "Look at the orchids!" The plants they see are not orchids, however. They are bromeliads; the name is derived from their botanical family name, *Bromeliaceae.*

Bromeliads grow mainly in tropical America, and most of them on host trees, although a few varieties grow on the ground. The pineapple is the best-known bromeliad and that is why the epiphytic members of the family are sometimes called wild pines. The Spanish moss that is so widespread through the Southeastern coastal plain is another member of this group; in general appearance it is the least characteristic of the lot. In the United States you don't get into real bromeliad country until you reach southern Florida. Some of the swamps, heads and domes of Big Cypress offer the best examples of epiphytes to be seen in the North Temperate Zone, and for sheer abundance the bromeliads there are as impressive as can be found anywhere.

The nearest thing to a bromeliad census in Big Cypress that I have heard of was a casual estimate made by three botanist colleagues of mine from the University of Florida: Dan Ward, John Beckner and John Carmichael. A few years ago they spent a morning wading around in the Fahkahatchee Strand. At noontime the three men stopped for lunch in one of the watery Gothic antechambers of the place, and as they stood there belly deep, eating sandwiches, unable to find a stump to sit on, they fell to cogitating on how many epiphytes there really were up there on the trunks and limbs of the six big trees that formed the dome of the room they were in. The thought took hold, and they decided to try to make some sort of an estimate. Each of them made spot counts of selected cubes of space, then multiplied these by figures they thought approximated the total epiphyte-covered area on the ceiling and walls of the pool-floored room. It was not a very precise exercise, obviously, but it was a useful thing to do all the same. When Ward, Beckner and Carmichael pooled their counts, the totals ranged from 1,000 to 10,000 orchids, and from 10,000 to 100,000 bromeliads—all in that one small corner of the Fahkahatchee Strand.

The bromeliads do more, however, than give an exciting look to the landscape. The leaves of many of the species are broad and cupped at the bases, and collectively these make a rain-water reservoir that may last through a normal dry season. Besides tiding the air plants themselves over periods of drought, these tanks are lifesavers at such times for a great many kinds of small animals, which keep from drying out by living in the plants. Various kinds of frogs, salamanders, snakes, snails and lizards, and a whole host of insects, spiders and scorpions make use of this refuge. For many such creatures, the moisture trapped in the leaves of bromeliads is a nutritious soup comprising rain water, dust, rotting leaf matter, animal droppings, the juices of plants and the decomposing remains of various insects. Some of the bromeliads that do not store water nevertheless make excellent hiding places for insects, and thus provide full-time foraging for such insectivorous birds as the blue-gray gnatcatcher and the yellow-throated warbler.

The heartland of Big Cypress—nowadays, at least—is the Fahkahatchee Strand. The name is a word from the language of the Miccosukee, one of the remaining tribes of the Florida Indians; it means forked river, though the two branches the Indians had in mind are no longer easy to locate. The Fahkahatchee, a 100,000-acre strip of wilderness some 25 miles long and seven miles wide, includes samples of almost every landscape to be found anywhere in Big Cypress Swamp. The Strand is a self-contained drainage basin that collects and distributes the rainfall of a great area. There are permanent lagoons and ponds in its interior, and during the wet season these spread out into a shallow, slow-moving watercourse that delivers vital fresh water to the mangrove estuaries of the Gulf Coast.

The deeper central parts of the Fahkahatchee once contained the best of Big Cypress timber. In the 1940s and '50s, however, a complex system of timbering trails and railroad lines was run out into all but a few small scraps of the tract, and wherever these reached, the mature timber was all removed.

The timbering of the Fahkahatchee prompted an all-is-lost attitude among most conservationists, and until lately this has hidden the fact that the place is still a treasure house of wild country. Much of the Strand has been drastically modified by man, but the place has not by any means been irrevocably ruined. It is still a vital asylum for harried wildlife and the home of such endangered species as the Everglades mink, the wood stork and the Florida panther. It also exerts a stabilizing hydrologic influence on the whole surrounding region.

I can remember my own feeling when they cut down the big timber in the Strand. I did just what I am complaining about in other people: I wrote the place off and for years made no effort to visit it again. But one day not long ago I drove out the Janes Road, the old timber road that runs northwest from Copeland, and I met a black bear and saw the royal-palm strand, and I realized how simple-minded it is to think that only virgin landscapes are worth saving.

It was the palm strand I was mainly looking for that day. I had spent the afternoon before in a little Cessna airplane flying over and up and down it. I was astonished at what I saw and decided I had to see what the strand looked like from the ground.

I had grown up believing that the only native royal palms in Florida were the seven or so that used to stand in Royal Palm State Park—now part of Collier Seminole State Park, near the western end of the Tamiami Trail—and a few others down on Paradise Key in Everglades National Park. Even those few palms, I had heard it said, had probably been planted by Indians. I knew vaguely that in the late-18th Century there were royal-palm groves in Florida as far north as the Saint Johns River, but I thought they had all fallen before some cold wave. So the royal palm had always seemed to me an exotic kind of tree, imported mainly, I supposed, by Thomas Edison and his friends to ornament Fort Myers, his winter home for half a century. Actually, as it turned out, most of the royal palms that have been used to decorate city parks and streets came from native Florida stock.

The royal palm—which no doubt got its name from its magisterial form and bearing—is one of the handsomest of all palm trees. It is stately to the point of stylization, and hardly seems to be a genuine vegetable. The trunk is like a concrete column, gray white, clean and unswervingly erect for up to a hundred feet. It holds up a heavy crown of long, feathery, gracefully curved dark green leaves, and in fruiting time a ponderous cluster of little nuts hangs just below the crown.

In southern Florida, the royal palm has been severely diminished in recent years because of the draining of its natural habitats. Those groves that withstood the frenetic draining and burning of the land have been steadily raided for young trees for nurseries and urban landscapes. For all that, the fact never got noised about that there were still hundreds of big royal palms and thousands of seedlings and young trees standing in the heart of the Fahkahatchee. Although local people and some outside deer, turkey and orchid hunters have known all along of the existence of the palm strand, I myself did not learn until lately that the remains of one of the most extraordinary plant communities in North America—a hammock swamp of bald cypress and royal palm—could still be seen there. It is perhaps the only undisturbed remnant of an association of the two kinds of trees that used to cover miles of the Fahkahatchee Strand.

The day I went out to see the place it was cloudy. Having no compass, I spent a couple of hours scaring myself by wading out into the swamp until I began to feel uneasy about which way would take me back out to the narrow road. I never had much sense of direction anyway, and the thought of how far it was through the strand to any other way home kept turning me back toward the road. But in spite of this I got far enough in from the road to see the big palms stringing out in both directions. Since then I have returned and worked my way farther back inside at several points. You have to wade to get in, and there is tangled country to claw through for a while before you reach the grove, but it is worth the trouble. If you keep looking up through the breaks among the crowns of the young cypress trees, you can make out the huge tops of the majestic palms towering high over the rest of the forest, and this way you can stay in the grove.

The whole interior is wet during most of the year. The water is mostly ankle deep to calf deep, though if you blunder into the pools and gator holes you can soak yourself to the ears. But this is a small price to pay for a walk among Florida royal palms. There are places in the strand that probably have no equal for their abundance of ghost orchids, swamp lilies, and bird's-nest ferns. Wherever the forest opens over pools and ponds, the limbs and trunks of the surrounding trees are wholly hidden by epiphytes. The buttressed stumps of the old big cypresses have moldered and crumbled into mounds of moist punk, covered over with great clumps of royal ferns. Here and there the grotesque basketwork of a strangler fig still squeezes the disintegrating body of an ancient cypress that was shattered by the falling of its fellows in the timbering days. Vanilla orchids, grass ferns and low-spreading peperomias make parts of the palm forest resemble much more tropical areas, reminding me particularly of the Caribbean.

The old royal palms are thinly spaced, but by searching out their tops through the occasional breaks in the cypress canopy you can get an idea of the density of the stand. Seedlings and young palms occur all through the forest; the older palms evidently suffered no ecological setback when the bald cypresses with which they shared the ground were taken out by the lumbermen. There are no old relict trees left standing as there usually are in cleared cypress swamps. When the Fahkahatchee was timbered, the logs were snaked out at high speed by cable, and even the old hollow or crooked cypresses—useless for logs—were cut down to clear the way. But the smaller cypresses and saplings that survived the violence are 30 years older now, and you can see what the place will be when, in time to come, their crowns have climbed up level with the crowns of the royal palms.

I also mentioned seeing a black bear in the Fahkahatchee Strand. He was just standing there in the road when I saw him; but bears are scarce nowadays in southern Florida, and you can travel through a lot of back country without seeing one. Besides, running into one that day was a curious coincidence, because I was looking for panther tracks at the time. After leaving the royal-palm strand, I had driven on out the

Leafless in dry seasons, dwarf-cypress trees support clusters of stiff-leaved wild-pine—an air plant related to the pineapple.

Janes Road to where it was blocked by the outermost canal of a ghastly gridwork of ditches that real-estate developers have dug out there in the wilderness. I stopped at the new gash, sat on the excavated limestone bank, and watched the clean, dark water washing out westward toward the sea, racing away like blood out of a big cut artery. I thought of the futility of the Fahkahatchee trying to heal itself in the face of such hemorrhaging, and got depressed beyond description at the sight. Having just come out of the royal-palm strand, where the regeneration of an incomparable landscape depends on a steady water cycle, I was forcibly struck by the senselessness of the canal; it seemed not just a lesion in the body of the land, but a public tragedy.

I got up off the rocks, still heavy-hearted, and stayed that way for a while as I drove back down the lonely little road. But then I saw some buzzards circling low over a shrunken water hole on the marl and remembered that I had intended to look around for panther tracks. The palm strand was the main reason I was out there, but I also wanted to try to find where a panther had walked. Practically all of the panthers left in the eastern United States are in southern Florida, not far from where I was, a short way to the northeast, or down in sections of Everglades National Park, where abundant deer and coons provide food and poachers are few. But panthers wander; I had heard that a deer hunter had seen one along a trail just off the Janes Road only a few weeks before. Though I have lived in Florida practically forever, I had never seen a panther here. What made this worse was that my wife, who grew up on the edge of Big Cypress Swamp, once saw one in the road behind her house. That made the gap in my life list of Florida beasts a very serious thing indeed.

It was the dry season. The pools and ditches were mostly just bare marl and muck and the ground was in good shape to take footprints. So when I reached a big spread of wet prairie I spent an hour or so walking out across it and on into the swamp beyond, looking hopefully about the marl for big cat tracks. Though it was pretty much a needle-in-a-haystack search, panthers are restless and leave a lot of tracks in a night's prospecting; if one has been around, and there is bare mud where he passed, you are likely to see his sign. But this time, in an hour of searching in the swamp, I found no big, round, clawless tracks at all. There were opossum prints galore, and coons had walked all over the place. There was also a lot of armadillo sign—triangular prints with a tail mark between them, like the spoor of some little kind of leftover dinosaur; and there were places where alligators and turtles had crawled over the mud from one shrinking pool to another. Herons had made cuneiform inscriptions at the edges of most of the ponds, and around one dried-down hole that was dismal with dead and bloated garfish, buzzards and an otter had turned the marl into a regular Rosetta stone of marks and scratches.

But no panther had been there anywhere. After a while I gave up the search and walked out to the dim timbering trail I had clung close to after leaving the prairie. It was there that I saw the bear. It was in the trail, looking the other way, and I quickly stepped behind a bush before it saw me. It shuffled back and forth beside a shrunken water hole in the lime rock, walked out into the water briefly, then went back into the trail and paced back and forth a few more times. It seemed undecided about something. That was about all I could gather, watching it —that it was probably worried about something. Then all of a sudden it loped straight into the brush and disappeared; I heard a stick snap as the willows took it in.

I walked to the car and drove to the Copeland fire tower to talk with the foresters about bears and panthers. They said both were out there, and that a few of each were seen every deer season, but there was no way to find either one on call. They said I ought to feel pretty good over just having seen a bear. And I was happy about it, of course; though you see so many bears in national parks that it somewhat dims the exhilaration of seeing one, even where bears are rare.

If the Fahkahatchee Strand is a good example of a partly sullied wilderness that is still well worth saving, Corkscrew Swamp, up in the northwest corner of Big Cypress, is the most important remaining fragment of virgin full-sized bald-cypress forest in southern Florida—a superb specimen of an almost lost kind of terrain. It ought to be seen by anyone who dreams of the world as it was before the arrival of man.

Corkscrew is now a sanctuary managed by the National Audubon Society. The features that attracted the support necessary to save such a place were mainly two: the stand of gigantic bald-cypress trees and the biggest wood stork rookery left in Florida. Both common egrets and wood storks have nested there in considerable numbers since early times. As long ago as 1912, when almost all the plume birds had been shot out of Florida, the National Audubon Society hired a warden to guard the Corkscrew rookery. The first land parcel in the series that produced the present sanctuary was acquired in 1954, after concerted efforts were made by various people and organizations, with the whole-

Air plants obscure the branches of pond cypresses and other host trees in the Fahkahatchee Strand. The giant wild-pines—those that look like pineapples—are the largest of the Everglades air plants; they grow up to four feet wide and six feet high.

hearted cooperation of the owners of the tract. Today the protected area, including its buffer zones of prairie and pineland, comprises about 11,000 acres. Development of Corkscrew Swamp Sanctuary must certainly be counted among the more outstanding achievements in the field of conservation.

You can walk out into the heart of Corkscrew on a 5,800-foot boardwalk, carefully laid out to take an observer through the different biological communities in the tract. Before reaching the swamp you go through a marginal zone of virgin slash pine, which is itself an extraordinary relic. This is kept free of encroaching hardwoods by controlled burning, and has the nostalgic look of a southern Florida landscape that was once widespread but now has almost completely disappeared. It is also the winter home of goldfinches, red-cockaded woodpeckers and brown-headed nuthatches. From the zone of slash pine the boardwalk cuts across a wet prairie and goes through a dense stand of pond-cypress trees bristling with epiphytes. After passing some little ponds, the walk enters the bald-cypress forest, with its understory of custard apple, pop ash, coastal-plain willow, swamp fern and fire flags. It finally ends, more than a mile from its beginning, in a broad central marsh around which most of the numerous nesting groups that make up Corkscrew's wood stork colony usually locate.

By all odds the most awesome of the Corkscrew communities is the stand of immense bald-cypress trees rising up out of gleaming black water. Some of the trees are more than 700 years old, and their massive trunks hold up heavy, flat crowns. The trees appear too short for their great girth, often as much as 25 feet—no doubt they were truncated by Hurricane Donna in 1960, or by earlier and even more fiendish hurricanes that all old southern Florida trees have weathered. What must be another sign that this is a hurricane community is the wide spacing of the cypress crowns. The sunshine that streams in among them has generated a prolific lower story of custard apple, pop ash, oak and maple trees, and beneath these is a welter of ferns and water plants. Though bald cypress is a less attractive host to epiphytes than pond cypress, big wild pines perch in crotches of the old trees, and butterfly and night-blooming orchids are numerous. Some of the cypresses are strapped and bound with strangler fig.

In this big-timber section of the sanctuary you may be disappointed to find no constant slithering and sloshing of wild creatures. As in most heavy woods, the inhabitants stay strangely quiet or out of sight. But you can be sure they are there, awaiting the right time of day or season, or a quieter and more patient watcher. Meanwhile, the botanical architecture of the place is a stupendous thing to see.

The value of an asset such as the Corkscrew Sanctuary is beyond calculation. Material values can be assigned to some of its aspects, but there can be no adequate assessment of the ultimate worth that is generated when, say, a lot of people just down from Michigan or Connecticut stand in the middle of a virgin cypress swamp on a winter day and hear an alligator bellow.

I saw that happen not long ago. It was a cold February day turned suddenly fair after high winds and rain the night before. The heavy cypress tops were thrashing under a gusty northwest wind that came in from the Gulf. Here inside the swamp it was too cold for the comfort of reptiles, and the people I met as I went along the boardwalk were mostly trying to name warblers, or just standing and pondering the Mesozoic look of the woods.

Then I came to a place where the boardwalk emerged from the swamp and cut across a small body of open water known as Lettuce Lake—a black-water pond carpeted with water lettuce, floating ferns and duckweed, and ringed about with willow, ash, and custard-apple trees. Out on the walk, in the sun-flooded midsection of the pond, some two dozen people stood. They were in three separate groups, and were gazing or pointing cameras at three different denizens of the swamp that were out there making the most of the morning sunshine. One was an alligator, an adolescent just under four feet long. Its head and shoulders were raised high above the floating lettuce, and though the rest of it was hidden, it obviously was standing on a submerged snag that gave support to its effort to rise as high as it could toward the comfort of the sun on a chilly day. This was an unusual pose for an alligator—a melodramatic one, in fact. Most of the viewers weren't necessarily aware of that, but they did volubly appreciate the large amount of this alligator that was out in view.

Another group of people were leaning anxiously over the rail on the other side of the boardwalk. Most of them held cameras or binoculars hopefully poised. All were looking down into a brushy tangle on a raft of root-laced bottom muck that had floated to the surface and now supported growing plants. I walked over and looked the way they were looking and gradually made out the dark brown, speckled, almost wholly camouflaged form of a limpkin, a long-legged, long-billed wading bird that is one of the essential creatures of the Florida swamps. Oblivious of its audience, it was standing on the raft and industriously probing for snails in the vegetation along the edges. For a while it poked around there, nearly hidden among the lizard tail and Indian turnip plants. Then suddenly it moved out from behind the screening leaves and the people could see it clearly in the viewfinders of their cameras; they all started furiously snapping pictures of the limpkin in case it should go away again.

I was only sorry it did no shouting. The voice of the limpkin—an unsettling sort of wailing cry—is wild and strange, and is bound to confirm an outlander's oddest notions about tropical swamps. But the limpkin is great to see even when it is quiet.

The people in the third group on the boardwalk over Lettuce Lake were watching the antics of an anhinga, or snakebird, a mature male in exemplary black-and-silver plumage that he was busily preening back into shape after the wind and wet of the night before. His wings were akimbo, his tail jutted to one side, his serpentine neck was curved in a long loop that brought his beak to bear on some spot of disorder between his shoulder blades. He looked like a product of somebody's first lesson in taxidermy.

Anhingas are weird, anyway. They are nearly three feet long from bill tip to tail tip, and are very thin. They have inadequate oil glands, and so have to spend a lot of time drying out after wettings, with their wings spread like an emblem in heraldry. In the water a snakebird will swim around under the surface stalking fish, once in a while slipping a foot of neck out above the surface and cruising about like a needle-nosed snake—or, if the fishing is good, coming up with a quarter-pound bream and juggling and tossing it in the air until it gets the headfirst grip it needs for swallowing. But an anhinga doesn't need to be hungry to go for a swim. When a snakebird tires of a perch it will as often as not just crash headlong into the water and disappear, perhaps showing up again far across the pond. When it has got so wet it can't fly, it will plunge out of the water like a sea lion and claw and chin its way up a post or tree trunk and hang itself out to dry in the sun.

The snakebird is really strange. It looks peculiar and acts the same. It can soar like a buzzard, and its voice—reserved mainly for the expression of resentment—is a coarse *buzz-buzz-buzz-buzz,* like a rotating rattle with gaps in it. And it seems to me that snakebird nestlings must be about the ugliest of animals.

The snakebird at Lettuce Lake was perched on a low limb, no more than eight feet above and a little to one side of the walk where the people were. So, being exotically photogenic anyway, and cavorting out there in the full 11 o'clock sunlight, he brought gratification to the aspiring wildlife photographers in the crowd. A dozen or more cameras were turned on him—from unassuming automatics to an elegant instrument with just the right telephoto lens to fill a 35-mm. slide with the blood-red eye of a breeding male snakebird. There was even a lady sitting out there in the middle of the walk with a short tripod in front of her and a tiny camera attached to an expensive and powerful miniature telescope. She couldn't have been more than 30 feet away from her subject. I judge she must now have some excellent portraits of the kind of feather lice that snakebirds have.

So it was a pretty satisfactory situation out there on the Corkscrew boardwalk; and all of a sudden it got astonishingly better. There was a rumble of blasting from some construction work far out in the northeast somewhere, and instantly an alligator answered it with a belly-deep roar from a patch of little willows not 20 yards from the walk. I looked that way and behind the willow switches I could see the old beast sloshing into a new position in six inches of water and mud. As I watched, it stopped and swelled and roared again.

Then I noticed that 20 or 30 baby alligators were swarming around and over the big one, and I realized it was a female, probably roaring defiance because she took the noise of the blasting for a neighbor that might have a mind to move into her nursery. Anyway she roared back at the now quiet dynamite six times in all, so close that the boards of the walk trembled beneath our feet. Before she stopped making that incredible noise, I lost all decent restraint and, to my sons' embarrassment, started rushing up and down the boardwalk, pleading with the puzzled tourists to appreciate how blessed they were.

3/ Gator Holes and Fish Jubilees

The whole system was like a set of scales on which the sun and the rains, the winds, the hurricanes, and the dewfalls, were balanced so that the life of the vast grass and all its... forms were kept secure.

MARJORY STONEMAN DOUGLAS/ *THE EVERGLADES*

Although the Everglades are fundamentally watery, the most relentless factor there is drought. Each year legions of small animals, and others not so small, dry up or suffocate during the winter dry season. Little rain falls in southern Florida from November to April, and the dying off is so widespread at this time it is sometimes hard to see how the animal life will revive with the May-October wet season. Some of the more wide-ranging animals—birds, mammals and flying insects—escape death by leaving the Everglades. Others go into a more or less torpid state called estivation, after walling themselves off in chambers or capsules that stay damp under the crust of sun-baked muck or marl. One salamander, the long, two-armed siren, makes a spherical chamber in the hardening mud and lines it with slime; here it can outlast drought that is not too protracted. In the cracked bottoms of vanished pools I have found congo eels—another type of long salamander—as well as live bullfrogs and four kinds of hard-shelled turtles.

Some creatures are able to find refuge in scattered ponds, holes and depressions that retain water even after the River of Grass has dried up in its bed. Crayfish and alligators dig their own water holes, and in the dry times these come to be occupied by a great variety of refugees from the drought. Crayfish burrows sometimes go down two feet or more, and many smaller animals join the crayfish in the moist gloom of their depths. Alligators gnash and slosh out pools—called gator holes—in the muck or marl, and these regularly turn into teeming little microcosms, where most of the aquatic creatures of the region find asylum. Besides these gator holes, alligators often make water-filled dens or dig caves back into the side of their gator holes, and in these they can sometimes survive baking drought even if the gator holes dry up. I know of one 10-foot live alligator that was dug from a mud-covered den in which it had apparently been lying for months after the bayhead it inhabited was drained.

But alligators do much more for the Everglades than just preserve themselves underground through times of drought. The area around the little ponds in which they pursue their more active existence serves an immensely important use for other creatures. In May or June, the female alligators build their nests—mounds of plant debris mixed with mud, and perhaps three or four feet high and six to eight feet long—usually located near a gator hole. Over the years the dredging up of muck and trash out of the ponds and the heaping up of nest mounds have combined to produce a unique feature of the terrain: a deep little pool flanked by a curved mound on which willow, buttonbush, myrtle and other small trees can grow. These mounds are often the highest ground for hundreds of acres around. Turtles lay their eggs on them, swamp rabbits and raccoons bivouac there, and birds nest in the trees and bushes. In the pools themselves representatives of most kinds of aquatic animals of the Everglades survive each year, and from these the River of Grass is repopulated when the spring and summer rains restore it. So a gator hole is far more than merely an asylum for the alligator that builds it; it is a major factor in keeping the Everglades environment both stable and diverse.

When the rainy season arrives in late spring the refugees in gator holes, crayfish burrows, or holes and cracks in the bedrock move out again into the newly flooded plain. Those that passed the dry time estivating underground push up out of the newly softened mud. Bacteria, protozoa and tiny crustaceans come out of their drought-resistant eggs or cysts; teeming new green algae begin their production of sugar to fuel the complex food chains of the reviving community. As fast as these smaller forms of life regain their abundance, the flying aquatic insects return, and the little fish appear, at first in small posses and then in bigger schools. Snails, glass shrimp, crayfish, turtles, garfish, mud eels and catfish venture forth from their various shelters. Aquatic plants sprout and grow. Bullfrogs and tree frogs sing, mate and lay eggs that quickly produce tadpoles. Half a dozen kinds of turtles climb out onto the few high places and leave their eggs to hatch wherever the marl or peat rises above the lap of the River of Grass. Even the bass and bream come back—from where it is hard to tell, since they are among the first to die and float belly up when water levels fall in winter. The alligators reestablish their territorial patterns, bellow, breed and build new nests. The water birds disperse throughout the reviving landscape. It is an essential pattern, and probably has been since the Ice Ages.

A mounded alligator nest—about eight feet across and two to three feet high —conceals a clutch of 30 to 60 eggs under its sawgrass cover. When the young are ready to hatch after nine weeks' incubation, they signal by grunting, and their mother, waiting nearby, rips off the cover of the nest so they can come out (overleaf).

Although drought has always brought recurrent stress to the Everglades, nowadays the normal peril is increased by a growing water famine. Ever since the first canals were dug over a half century ago—in order to drain the water off the incredibly rich muckland soil and realize the golden promise of winter vegetables for the New York market —water levels in the Everglades have been falling. The dry seasons have become longer and more pronounced, and catastrophic drought and the fires that follow occur more frequently. On the other hand, not all the ecological problems of southern Florida are caused by a scarcity of water. There can be too much water, when successive hurricanes bring unusually heavy or unseasonal rains, or when the managers of the impounded conservation areas try to relieve their own flood problems by opening sluice gates and sending vast and sudden surges of water into the downstream Glades when they are already saturated.

The essential point, in short, is that the health of the region suffers when any phase of the annual wet-dry cycle—the seasonal balance of drought and flood—gets out of gear, for whatever reason. The original animals and plants of southern Florida were there not simply because the place was wet, and certainly not because it was dry, but because a productive high-water period regularly alternated with a time of drying down. This annual cycle has shaped the whole ecological organization of the region, and it molds the life cycle of every animal there.

There is no more graphic example of this balance than that of the reproductive cycle of the wood ibis, or wood stork, known among old Floridians as the ironhead. By any name, this bird epitomizes primeval Florida to me. As far as I am concerned ironheads are esthetically indispensable. They fairly exude atmosphere. The most vivid recollection I have of my first trip to the Everglades, years ago, is of six of them standing in serious caucus in a little glade set with clumps of slender palms that were silhouetted against the mist of an early morning. Ever since, the Everglades have been an enchanted place for me.

Ironheads are white and black and half as tall as a man. They have long meatless legs, a bald knob of a head, bare cheeks and a bill 10 inches long, three inches deep at the base and a little curved at the tip. It is the bald head and broad base of the beak from which the "iron" in the local name derives. The birds clap their beaks at you, and the way they use them in their fishing is a marvel. Ironheads eat small fish. They hunt by moving along in shallow water, crouched over forward, touching the tips of their partly open beaks against the bottom. As they walk they usually hold one wing nervously stretched out to the side—to scare up prey, perhaps—and they shuffle along that way, pushing first one foot and then the other jerkily forward as if to muddy the water and herd small forms of life into the trap of their beaks. Wood storks hunt by touch rather than by sight, no doubt because the touch reflex is faster than the interplay between eye, brain and beak muscles that visual hunting requires. You rarely see a stork visually spotting and selecting an individual victim to grab. Instead, it gropes blindly along in the shallows with its beak partly open, snapping reflexively when it makes contact with a fish. The snap reflex is incredibly fast. In tests with a captive stork, the zoologist Philip Kahl measured the time between a touch and the closing of the beak and found it to be only about 1/25,000 of a second. He also proved experimentally that successful snaps do not depend on sight; when he covered the stork's eyes with blackened halves of Ping-Pong balls, its ability to fish was not impaired.

For all its remarkable fishing skill, this bird is perilously close to extinction. The whole future of the species may depend on what happens to the few protected nesting colonies that remain in Florida. You sometimes hear people say what the hell—there are plenty of wood storks in Mexico; but the last time I looked there I saw a disheartening number of them smoked and stacked for sale in the Veracruz market. From what I see and hear, there is no real comfort to be found anywhere in the ironhead situation. Certainly, throughout southern Florida, it hangs in critical balance.

One factor in the decline of wood-stork populations has been man's manipulation of drainage and the interruption of overland flow of water by dams and dikes; this interference retards multiplication of the little animals the birds feed on. Another factor is the disruption of the balanced alternation of high-water levels that produce food supplies with the low levels that concentrate food during normal dry seasons. Ironheads are so sensitive to this balance that whenever the wet-dry

An alligator hatchling—only minutes old and about nine inches long—crawls over its still unhatched siblings' eggs in search of its mother.

cycle is upset, for whatever reason, their reproduction is imperiled.

Ironheads in the Everglades begin their breeding in November and December, early in the dry season. One recent November day I visited Corkscrew Swamp Sanctuary to see ironheads nesting there. Directly over a section of the boardwalk in the interior bald-cypress forest there was a cluster of 30 or more nests among the branches of the big trees. Some of the nests were still under construction, and others no doubt had eggs in them. There was no sign of young, but there were a lot of wood storks sitting up there, perched hunched-shouldered and quiet or fussing over nests or clapping their bills at the people staring up at them. Out around the edges of the broad Central Marsh there were more clusters of white in the cypress tops, where other sections of the colony had lodged. A ranger I met told me that there were still other nesting groups, and that an aerial survey the day before had tallied about 4,000 pairs.

Two weeks later I went back to see how the rookery was getting along. I expected to see a lot of nestlings by then. To my surprise I found the nests over the boardwalk completely untended, with no storks to be seen anywhere. I scanned nest after nest through my binoculars and finally made out a patch of white where one lone bird was setting. But the colony had obviously been abandoned. Even out around Central Marsh, the six or eight nesting colonies that had been there two weeks before had disappeared. Wherever any spot of white showed, it was only one lonesome-looking bird.

The change was dramatic and puzzling. I had long been accustomed to associating the ecological troubles of southern Florida with lack of water. Though the rainy season just past had been a little drier than normal, it had been no worse than others in which up to 6,000 pairs of storks had nested in the sanctuary. I asked a ranger what had happened.

"Rain," he said.

"What do you mean, rain?" I said, hung up on the idea that Everglades troubles come from *lack* of rain.

"Sure," the ranger said. "Too damn much. It rained so much the storks figured they better not try to breed after all."

The ranger's remark made no sense at the time. But later, when I thought about it, the reason for the Corkscrew catastrophe became clear. I remembered what I had read about the dependence of storks' nesting success on the normal wet-dry cycle in Kahl's paper on the food requirements of the wood stork in Florida. According to Kahl, the bird usually nests at the outset of the dry season for two closely interrelated reasons: first, the recent wet season has brought populations of the small food fish that are preferred by the stork to their yearly peaks; second, with the falling water levels of the dry season, the forage fish are crowded closely together in diminishing ponds and pools, and the storks can catch them in the great numbers needed for feeding nestlings. The shrinking of the ponds and the consequent concentration of the food supply apparently generate the decision to nest. Florida ironheads, and perhaps ironheads everywhere, have developed an ability to judge when water has fallen low enough that nestlings can be amply fed. If the proper lowering of levels fails to materialize, the storks simply refuse to breed. Occasionally, if breeding has already begun and an unseasonal rainfall occurs, the rain will abort the breeding. That is what happened in the case of the nesting failure in the Corkscrew rookery. The storks evidently abandoned their nests because they had been discouraged by a few days of heavy rain in what was normally the dry season.

For a bird that fishes by random groping and touch, anything that reduces the density of the fish supply—as rising waters do—is a setback. At nesting time, when wood storks require a prodigious amount of food, it is a calamity. Kahl estimates that a family of storks, including two young ones, needs about 440 pounds of fish to sustain itself during the four months of the breeding season. Over that period a colony of 6,000 pairs—roughly the average population of the Corkscrew ironhead colony—would require more than two and one half million pounds to get by. And most of this food comes in small packages. While the stork catches many different kinds of small fish and even some lizards, snakes and frogs, its most frequent prey is the mosquito fish.

Compounding the problem is the particular way in which storks handle their parental duties. According to Kahl: "Both parents feed the young. Food is carried in the throat or stomach of the parent and regurgitated on the nest floor, where it is picked up by the nestlings. Up until the age of three or four weeks, nestling storks are seldom left unattended; one parent remains at the nest at all times while the other parent forages. . . . By three or four weeks of age the nestlings are able to defend themselves and both parents begin to forage at the same time. The period before the young are left alone is especially critical since the increasing food demands of the nestlings must be supplied by the fishing efforts of one parent."

Obviously, storks have to work hard to meet these demands. To locate fishing holes, as well as to extend the limits of their daily hunting travels, they augment their own powers of flight by using rising currents of warm air—as many birds of prey do. If on a clear, calm morning you find yourself anywhere near an ironhead rookery, you will almost surely see storks spiraling over the nesting trees, climbing in the updrafts that the morning sun generates when it heats the land. By riding these thermals, a bird can climb high without spending much energy, and in that way can vastly extend both its field of search and its foraging range. If their nesting period is to be successful wood storks are compelled to make daily trips in search of some densely populated, drought-shrunken pool. What determines a good hunting ground for them is not fish per square mile, but fish per heavily crowded cubic foot of water. Because they depend upon prey randomly colliding with their beaks, and because the dependence is obviously most urgent when young are waiting to be fed, they may have to travel farther than they can fly under their own power to find the right kind of fishing hole.

So if you go to a southern Florida wood-stork rookery on a morning when a ravenous new generation of nestlings has emerged, you will surely see the parent birds flapping from the tops of the mangroves or cypresses in which they are nesting. They will rise in growing circles as they move out over the sawgrass; then, climbing as high as the buzzards, they will either move across country with the updraft or glide out to a good gator hole they already know, or perhaps to a new one they have spotted from their high vantage point.

In front of my home in northern Florida there is a 10-acre pond. Every decade or so the bottom drops out of a sinkhole at one end of the pond bed and the place goes nearly dry. Although the pond is hidden by trees and is five miles from any regular ironhead haunt, it takes no time for the storks to learn of the low water and gather at the pond for the easy fishing. But five miles is no distance for the storks to travel in their search for food. Kahl has found that storks sometimes commute as far as 30 miles each way between the Corkscrew rookeries and fishing holes up around Lake Okeechobee.

For most Everglades creatures, the periodic drought that is essential to the nesting and breeding pattern of the wood stork requires the single most important adjustment of their lives. There are two aspects to that adjustment. One is the capacity to resist suffocation and total dehydration in the dry season; the other is the ability to spread rapidly and recolonize new expanses of water as soon as the wet season returns.

Two wood storks wade in a shallow Florida pond to stir up mud and bring an ample supply of small fish within easy reach of their bills.

Gambusias and their various relatives in the killifish family—which includes some of the smallest fish in North America—are specialists in both skills. Some of their feats in withstanding the ordeal of the dry season—living for days almost without oxygen, packed together by thousands—seem nothing short of black magic. The last time I was in the Everglades during a dry April I walked out on a marl prairie south of Alligator Alley to a shallow prairie lake that had dried down. A water hole, no more than 20 feet across, was all that remained. The outermost rim of the pool was ringed with the corpses of bluegill bream, warmouth bass and largemouth bass. These were being picked over by buzzards. On the moist mud inside this ring there was a mosaic of dead garfish and smaller fish of a number of kinds. At the innermost center of the pond, where the water was still in a semifluid state, there was a closely packed mass of very small fish—all killifish: flagfish, golden top minnows, least mosquito fish, a few lucanias and myriad gambusias. Although nearly immobilized by crowding in their diminishing puddle, many of them were still alive.

I couldn't tell how long it had taken the pond to shrink to that size. But it must have been at least three or four days earlier that the bass and bream had died from suffocation, at a time when the area of water was many times greater than at this final melancholy stage. So for at least three days the fish I found still alive had managed to survive in a teeming body of noxious warm water that was surely devoid of oxygen. I remember thinking at the time that this represented some kind of paradox; the visual evidence in the pool simply did not add up. The fish still alive in that dismal soup were kinds that have no known means of supplementing their aquatic respiration by taking in air, as some species do. What, then, had kept them from expiring long before, when the last oxygen in the awful little place disappeared?

It was a real puzzle, and one with obvious bearing on the machinery of Everglades ecology. Later I came across what may be an answer. It appeared in a paper by William Lewis Jr. of the University of Georgia dealing with the ability of killfishes—both the live-bearing and the egg-laying kinds—to survive in oxygen-deficient waters. Dr. Lewis concluded that the tilted mouths and flat heads of these species enable them to tolerate such conditions by taking oxygen from the inexhaustible reservoir of dissolved gases up in the thin surface layer of water where it is in contact with the air.

I am ashamed that this reasoning never occurred to me. I had spent quite a lot of time pondering the question of how flagfish, golden top minnows and gambusias—all similarly tilt-mouthed and flat-headed—are so marvelously able to survive drought long enough to be found wiggling in a hatful of wet mud, when this is all that remains of a 20-acre pond. Everybody has seen fish "gasping for air" at the surface of an undersized pool or container. Until I read Lewis' paper it was never really clear to me that the uppermost surface layer in any body of water, no matter how polluted—an aquarium, a bucket of bait or an Everglades water hole—is never completely exhausted of oxygen. In this layer oxygen is moving in from the air as fast as it is absorbed by the body of the water below. It is this thin stratum that crowded fish gasp after. But the heads of bass and bream are high, their mouths are placed at the front of their heads; to keep their mouths at the surface they would have to swim around tilted at an angle of 45° or more. The upturned mouths of killifish, on the other hand, allow them to stay in a natural horizontal position while they systematically suck in the oxygen-rich upper layer of their habitat. Lewis believes that this adaptation could account for a great part of the extraordinary drought resistance of the mosquito fish and its relatives. If so, I suggest that it accounts for a large part of the ecological organization of the Everglades.

But obviously, the ability of a small fish to live through the dry season in an isolated, oxygenless water hole will not in itself contribute much to the ecological productivity of the River of Grass. It is just as important that the forage fish have the ability to recolonize their wet season habitat when it is restored by the coming of the rains. When the wet season returns it may last only a few months. If the expanse of new water is to support new populations of larger animals, the little fish they eat must be able to spread as fast as the water spreads.

When a period of drought in the Everglades is ended by rains, and residual pools join together to form shallow lakes, it is a marvel to see how fast the new habitats are repopulated. Some animals, such as birds and aquatic insects, simply fly in. Some come up out of estivation in the crusted-over mud, or hatch out of drought-resistant eggs or cysts left by progenitors before the dry season. But most Everglades fish have to swim to the revived country, and it is remarkable how promptly they return to it. The principal pioneer species, gambusia and its kin, are to some degree omnivorous; so presumably they have no trouble finding food in newly flooded territory. But what are the goads and signals that set them swimming out into new habitat and that tell them which direction to go in? They get about too promptly merely to be wandering aimlessly.

An important aspect of the pioneering ability of Florida pond fish and other kinds of aquatic animals may be their occasional custom of leaving quiet water and making mass upcurrent migrations when drought is ended suddenly by heavy rain. At these times you often can see fish, aquatic amphibians and crayfish that have gathered by hundreds, even thousands, in pools below stretches of fast water or other obstructions in temporary or swollen streams—at the downstream ends of culverts, for instance. At the University of Florida we have borrowed the term jubilee for these gatherings, taking it from the famous springtime congregations in Mobile Bay of salt-water fish and crustaceans, which are pushed by a layer of oxygen-depleted water toward the beach, where most of them die. The jubilees of peninsular Florida apparently have nothing to do with oxygen-depletion, and their goal is not death on the shore but new uncrowded space to live in. The only way I can account for these fantastic conclaves is to suppose that the individual members of these assemblages have been stimulated by a rain-generated current to use it as a road to a new habitat upstream somewhere. And obviously they are right—the current is a way to survival; any flow that comes into a crowded, dried-down water hole is a pathway to a new life for its inhabitants.

This jubilee behavior seems inconsistent at first because the fish, amphibians and invertebrates most frequently and abundantly involved are not current-loving species at all but inhabitants of quiet water. But when the new water flows into ponds and pools cramped by drought, the jubilee creatures suddenly become current seekers, presumably reacting to the flow because it bears the exciting taste of renewed opportunity upstream somewhere.

When you find one of these sporadic upstream migrations in progress, gambusias are nearly always involved, sometimes by tens of thousands in volumes of only a few cubic feet. In their numbers and density, gambusias are accomplished pioneers. They are also willing to eat almost anything nourishing, and they produce live young. In the mosquito fish, fertilization is internal. The female gives birth to a few fully formed offspring, which the moment they appear seem ready and eager to take up adult responsibilities. Thus, for the small creatures of the Everglades, disaster is built into the normal yearly cycle. But not being prone to anxiety they flourish, holding on in the gator holes during the dry time, going off on jubilees when the first spring floods come, and in summer scrambling back out through the reborn River of Grass.

If the wet season brings plenty of rain, the numbers of small fish that will crowd into the shrunken ponds and water holes when another dry season comes will be cause for celebration and conviviality among the larger animals of the Everglades—not just the ironhead storks but most of the other fish-eating creatures as well. A hundred years ago, an Everglades gator hole in the early dry season must have been a spectacular bit of landscape—not the equal of a good water hole in Tanzania, because the Everglades creatures are mostly smaller than those of Serengeti, though sometimes every bit as feverishly active. Even today, although the populations of most of the larger Everglades species are less numerous than they used to be, a gator hole in the early days of a dry season brings together in fantastic juxtaposition a wonderful array of predators and prey. When the River of Grass dries up, alligators and otters are crowded with their own kind, and with unusual concentrations of raccoons, herons, garfish, snakes and soft-shelled turtles. You can still sometimes see two dozen water snakes at once around the edges of a fish-filled gator hole, and maybe 10 kinds of wading birds will be there stumbling over the snakes.

So in the Everglades, each year brings a time of dwindling and bare survival followed by a time of spreading and explosive reproduction. Millions of individual animals die in the rhythmic disaster of winter drought; but the species to which they belong are adapted to disaster and will certainly revive again when the rains return, if man will only curb his abuses of the landscape they live in. Extinction has no natural place in Pahayokee. There is only an endless round of decimation and spectacular renewal.

The Ordeal of Drought

In the Everglades, the pendulum of the seasons swings in a wide arc between flood and drought, producing alternating periods of fecundity and barrenness. From early spring well into autumn, the Glades are usually kept flushed by ample rainfall, annually averaging about 50 inches. But by November the last drenching tropical storm of the year has gone, and its passage signals the start of the dry season, a time of trial for the creatures of the Glades.

The seasonal change takes effect slowly, after weeks of little or no rainfall. As the water level is lowered by evaporation and runoff, the higher terrain drains and the low, muddy areas dry in a maze of cracks *(right)*. The isolated ponds that remain in small depressions trap great numbers of fish in their dwindling waters. In the desiccated areas, the shallow-rooted grasses and weeds gradually turn brown and die off, providing fuel for the ground fires that normally burn out large tracts of vegetation. In years of extreme drought, even the wide sloughs have dried out completely or shrunk to streams too narrow to serve as effective firebreaks.

By late January as a rule, the lack of water and the related threat of famine confront the creatures of the Glades with a deepening crisis. Each species meets this challenge in its own way. Many birds fly off to wetter habitats. Many land animals—deer, raccoons, opossums, panthers and various small rodents—hold on, though they are forced to spend their days roaming the dried-out marshes in search of water.

Some species, unable to sustain normal activity, retreat into estivation, a torpid state similar to hibernation but not as deep or persistent. Estivating tree snails conserve their bodily moisture by attaching the openings of their shells to the bark with an airtight mucous seal. Apple snails, frogs and turtles dig into the moist undersoil and there lapse into a self-preserving lethargy. During especially severe droughts, even alligators will estivate in damp dens beside their shrunken water holes.

Sometime in May, usually, the first abundant rainfall in months begins to quicken the pace of life in the Glades. By then, the drought has served a grim but essential purpose. Many creatures of every species have died of thirst and hunger; and this natural check on their proliferation helps to ensure a stable, balanced population in the Glades.

Cracked soil and withered sawgrass testify to the drying out of water sources along the Shark River Slough. The slough itself, up to 40 miles wide during the rainy season, normally shrinks to a few narrow streams during the November-to-April dry season.

Dry Times in a Shallow Pond

As the water level recedes in the low, flat Glades, a few feet of elevation can make the difference between life and death for innumerable fish during the dry season. Many of those trapped in dwindling pools have little chance to survive. Yet as ever in nature, nothing is wasted. A fish that dies will make a meal for a ravenous raccoon, and the subsoil beneath even a desiccated pond will retain enough moisture to save the life of an estivating turtle.

In general, the first casualties in a shrinking pond are its largest fish, such as bass and bream. Sunfish and other medium-sized species are the next to perish.

As evaporation and overcrowding progressively reduce the oxygen supply in the pond, these fish cannot breathe; their mouths are too big and clumsy to utilize the last resource—a fragile film of relatively oxygen-rich water at the surface.

Some fish survive through special adaptations. The tiny mosquito fish has an uptilted mouth that enables it to take in oxygen while remaining horizontal in a mere trickle of water. The large garfish is equipped with a primitive air-breathing lung that sustains it up to a point—even out of water. And, the bowfin, similarly equipped with a lung, can burrow into the bottom of a completely waterless pond and breathe the meager air that filters through a thin layer of mud.

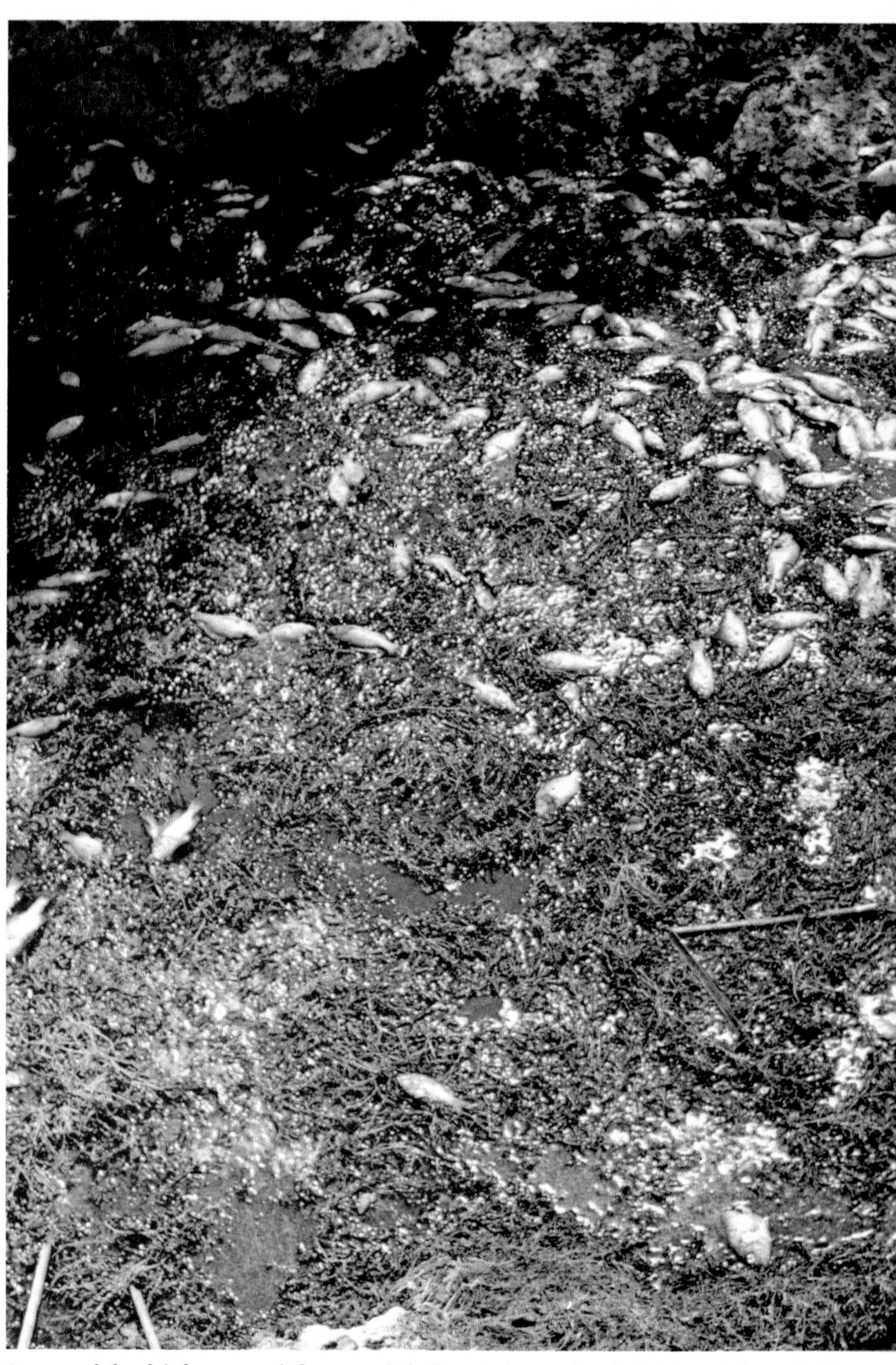

Scores of dead fish, most of them sunfish, lie victim to the shrinking of their shallow po

sh in deeper rivers and sloughs usually survive drought.

Predator or scavenger depending on opportunity, a coon devours a dead fish.

Preparing to estivate, a slider turtle digs into the moist soil under a dried pond.

Drying footprints trace an alligator's search for water.

The Gator Hole: Peaceful Refuge

Of the Everglades' many debts to the alligator, none is more important than the benefits that accrue from its relentless search for water in the dry season. Moving across the arid flats, the great reptile keeps its heavy body elevated to make maximum speed, thus leaving no belly marks between its footprints *(left)*. Sooner or later, instinct and persistence lead the alligator to water, or to a low place where the water table lies close to the surface.

The alligator then sets to work either to deepen and enlarge the existing water hole or to excavate a new one, breaking the caked earth with its powerful tail and shoveling away the debris with its broad snout. During the worst droughts, alligators have been known to dig their way down through four feet of compacted mud and peat before water has oozed up from the porous bedrock.

Such gator holes, found throughout the parched Glades, attract many thirsty creatures ranging from otters to herons, and soon become biological microcosms of the whole region. The alligators, conserving energy and living on their own fat, largely ignore the intruders; the refugees sustain life on the gator hole's remaining fish, insect life and vegetation, and live side by side in a relative state of truce. When the rains finally return, it is from these oases that the various species go forth to repopulate the Glades.

An alligator relaxes on the mudbank near its watery den. Several alligators—and many smaller animals—may occupy one gator hole.

4/ Sanctuary in the Park

The flooded forest, combining... the forces of earth and water, was a common ground where all creatures moved in quiet, with respect. PETER MATTHIESSEN/ *AT PLAY IN THE FIELDS OF THE LORD*

The squall had moved out of the drenched land an hour before and now was lingering offshore in the Gulf, its thunder dwindled to sullen rumbling in the night air. I had walked along the prairie road west of the Everglades National Park buildings at Flamingo, hoping the mosquitoes would stay hidden for a while. My flashlight revealed a little water snake on the road, and the eyes of a raccoon and two marsh rabbits. The thin trill of a cone-headed cricket came from somewhere in the grass, and along the shore a squawking great blue heron flew off into the darkness. At least I supposed it was a great blue. Cape Sable has great white herons too, and I wondered idly whether I was hearing one of them. Then a lightning flash printed a spread-winged image of the heron against the sky, and I saw it was a great blue after all.

In the quiet after the heron had gone I heard a lonesome-sounding frog chorus far off across the prairie. It was almost too far for me to sort out the voices in it, but I gradually made out the piteous *ma-a-anh* of narrow-mouthed toads, the yelping *erp-erp-erp* of the green tree frog, and a rasping chant that could have been either the *ark-ark-ark* of the Cuban tree frog or the flatter *ack-ack-ack* of the rain frog.

I was sleepy and didn't feel like sloshing across the wet marl to see for certain what other frog was singing with the green tree frogs and narrow-mouthed toads. But the singing frogs reminded me of the much bigger chorus I had heard up at Taylor Slough the night before and I thought, as I have many times, how neglected frog songs are. Not scientifically neglected—herpetologists spend quite a lot of time taping frog calls—but esthetically neglected. Frogs do for the night what birds do for the day: they give it a voice. And the voice is a varied and stirring one that ought to be better known. Ever since the days of the first talking pictures, Hollywood movie-makers have used recordings of the voice of the Pacific tree frog as background sound for scenes anywhere from New Guinea to the Carolinas. That they have been getting away with this decade after decade suggests to me that there has been a melancholy decline in the sensibilities of Americans. Frog songs are distinctive; every kind of frog has its own voice. You can learn to recognize frog songs as readily as the songs of warblers. Although frogs are fewer in species than birds, Florida is blessed with a great many kinds, and because they sing mainly at night when most birds are quiet they give the wet places much of their incomparable nighttime atmosphere. Perhaps the reason frog songs are not generally appreciated is that they are sung in places where mosquitoes and snakes live.

I still treasure the memory of a night when I stood by a pond with a little swamp at either end, and heard 14 different kinds of frogs singing all at once. That happened in northern Florida, where frogs are more diverse than anywhere else. You may never hear quite that many kinds in the Everglades, but you can easily find choruses of six or eight species in good frog weather. I recall a June night when the frogs drowned out the bellowing of an alligator I was trying to listen to.

Although frog-song is mostly confined to the night, it often breaks out on cloudy days, when the barometric pressure and the relative humidity are propitious. Two kinds of frogs can be heard in warm weather in the Glades on all but the brightest, driest days. One is the little cricket frog *Acris,* whose *ik-ik* is the most ubiquitous amphibian voice of the region. The other voice is the baritone *urr-urr* of *Rana grylio,* the southern bullfrog, which also is known as sulfur-belly, and pig frog. I have never liked the name pig frog, not because I have anything against pigs, but because *grylio* doesn't sound like one. The grunt of a pig has a nasal quality, a French *on*-sound; and it is usually two-toned, rising in pitch or else trailing off into a similarly Gallic *in*-sound. There is nothing nasal about the sulfur-belly's voice, nor is it two-toned. It is a deep, short, resonant, staccato trill, as if the *urr-urr* were being said by a person from some part of Spain, where "rr's" are reverently rolled.

The big bullfrog, *Rana catesbeiana,* lives in southern Florida too; but because it is partial to wooded ponds and tree swamps, its ponderous *jug-a-rum* is rarely heard in the Everglades where *grylio* holds forth. There is, however, another sort of bullfrog out there—if by bullfrog one means any member of the genus *Rana*. This is the slim, far-jumping southern leopard frog, which has a less authoritative voice than the other two. Its voice quavers or jerkily scratches out its song like a dry hand dragging across a tight-blown rubber balloon. The leopard frog is a superior animal, alert, sharp-faced, great at hurdling across broken country; but because its voice is both seasonal and quavery, and not too different from that of New England and Wisconsin leopard frogs, I lean toward the sulfur-belly as the true cantor of the River of Grass.

The Everglades are the sulfur-belly's kind of country. It is indispensable in Everglades ecology, harvesting tons of crayfish, aquatic insects and snails, and itself providing a staple food for reptiles, birds and bass. It zealously guards its territory. The drum-roll sound of its *urr-urr* is believed to serve as a signal for proclaiming its territorial rights.

The sulfur-belly has suffered grievously from market hunting. Along with the alligator and the plume birds, it stands out among the creatures that have borne the brunt of man's effort to force a living from the Everglades wilderness. The sulfur-belly is the main market frog of southern Florida, and when airboats began to ply the Everglades in the 1930s frog hunting quickly became a specialized calling. The biologist Frank Ligas has estimated that the best frog hunters used to get up to 400 pounds of legs in a two-night hunting trip. They can't match that today. Sulfur-bellies feed heavily on crayfish. The dry times have become too hard and long to keep crayfish abundant in the River of Grass. And the frog hunters themselves destroyed a lot of bullfrog habitat by hunting alligators, legally and illegally, on the side. As they killed out the gators, the gator holes and water trails grew over with vegetation, driving the frogs away. But down in the heart of the Everglades National Park, where all hunting is prohibited, the frogs are coming back.

As anyone would expect from looking at a map, the animals and plants that have colonized the relatively new land at the tip of the Florida peninsula are a mixture of temperate-American and tropical kinds. As also might be expected, species derived from the continent predominate, simply because salt water separates Florida and the West Indies, and most creatures have a hard time crossing salt water.

The animals and plants of new land that appears above the sea don't spring full-blown from the ground. They have to move in from other places. How they move and where they come from is the specialty of

A nimble climber, this green tree frog owes its agility to specialized feet, which have large toe pads that grip like suction cups.

biogeographers. Up in northern Florida, where I live, the land is maybe 15 million years old, and the animals and vegetation of northeastern North America have had plenty of time to colonize. But down in southern Florida some of the land was being formed only about 3,000 years ago. Because the dry ground there is separated from the nearby West Indies by salt water, and from the continent to the north by broad expanses of glades and prairie that vary from bone-dry to watery in a normal year, you would expect the fauna to have been built up out of the more vagile—the more biogeographically footloose—kinds of animals. And so they have been. The region is heavy in birds, for instance. In fact its crown jewels are its birds—tropical forms such as the black cuckoo-like ani, the white-crowned pigeon and the roseate spoonbill; the wide variety of resident species derived from the temperate zone; and the hosts of migratory warblers, thrushes, vireos, ducks and shore birds that stop over in their travels. Many wingless creatures that live with special sets of ecologic conditions—as earthworms, salamanders and upland fishes do—may have found it impossible to reach whatever small tracts of suitable habitat may exist in southern Florida.

The land platform on which Florida rests is very old, and though there have been repeated floodings and withdrawals of the sea during the last million years, there was land in central and northern Florida most of the time, and terrestrial life persisted there. It is from there that much of the fauna and flora of southern Florida was derived.

Nonaerial animals vary markedly in their ability to get across salt water. Lizards are good at island-hopping; salamanders are quite unventuresome. The capacity to colonize new range is not directly related to locomotor ability. Frogs, though able hoppers, are easily foiled by the barrier of salt water. One reason for this is the dual nature of their life cycle. As tadpoles, frogs resemble fish; when they mature they are terrestrial but scaleless, and prone to desiccation and salt-pickling. No frogs have come into southern Florida from the West Indies by natural means. The two tropical kinds that do occur—the Cuban tree frog and a tiny sweet-voiced greenhouse frog with the generic name *Eleutherodactylus*—were almost surely introduced accidentally by man.

The most conspicuous tropical element in southern Florida is botanical and it is not hard to see why. Most of the West Indian plants that occur there have fruits, seeds, or other reproductive bodies that are susceptible to distribution by storm winds or ocean currents. Some of these may have been brought in by birds. More than a hundred of the tropical southern Florida trees have small fruits that birds eat regularly. Anyone who has parked his car where migratory robins roosted after raiding a cherry-laurel tree will know what sowers of seeds a flock of birds can be. There is no doubt that within southern Florida birds carry the fruits of pigeon plum, saffron plum, coco plum, satinleaf, mastic, Jamaica dogwood, silver palm and a host of others, and set up new stands wherever the seedlings are able to take hold. So perhaps some tropical plants were brought to southern Florida by birds.

Whatever the origin of the Everglades flora and fauna, they have managed to adapt to a rigorous environment. At first glance the tip of peninsular Florida might seem like an easy place to get along in—jutting down almost into the tropics, nudging into a curve of the Gulf Stream, with freezing temperatures rare, and with up to 60 inches of rain a year. It sounds like lush, lazy country, in which any creature able to get there would surely find a way to live. But despite the absence of tumbling streams or grinding ice, the Everglades Basin animals and plants have developed under stress: drought, flooding, fire, the winds and driving water of hurricanes, even sporadic killing frosts. These factors—some regular and predictable, some operating only once in a decade or so—have surely kept out many potential colonists. The species that have successfully pioneered are, in one way or another, adapted to absorb the recurrent shocks and then recoup.

The most dramatic trials they face are the hurricanes. In 1960 Hurricane Donna drove tons of salt mud across the prairie, tore up ancient forests of mangroves, and left 200-pound loggerhead sea turtles slogging about in the marl far inland. After the hurricane of 1935 we went looking for crocodiles at Madeira Bay and found enormous masses of reinforced concrete, chunks weighing 10 tons or more, in the edge of the mangroves; they had been part of the Lower Matecumbe ferry slip, 30-odd miles across Florida Bay.

For all the ecological vagaries, the Everglades provide a refuge that may be vital to the survival of a number of kinds of animals that are declining seriously outside southern Florida. Most of the endangered or declining species are birds. It was in what now is the Everglades National Park that the most important episode in the hairbreadth rescue of the plume-birds occurred. The spirit that accomplished that unprecedented feat was of unreckoned significance in the evolution of human concern for natural species.

Few people are aware that in spite of the environmental disruptions of the past few decades, you can now see more herons and egrets in a day's drive through southern Florida than you would have seen in a whole year back in 1905. The change came about mainly through suppression of the feather trade. The tales of the killing that took place at the rookeries to adorn ladies' hats sound today like nightmares.

It was the establishment of the Audubon Societies and the courage of their early wardens that started the trend against this slaughter. The murder of an Audubon warden, Guy Bradley, by poachers at Flamingo in 1905 focused the indignant attention of the world on the bird plume industry, and soon thereafter the fad for the feathers began to languish. Since then there has been a creeping decline, even in the rural southeastern United States, in the idle shooting of herons. There has also been a blessed disappearance of the kind of hunter who would petulantly shoot any bird that came within range. Today, finally, such people are being looked on with more disgust and less indulgence. The culprits still exist, but nowadays increasing numbers of their fellows regard them as misfits, and this gets to them harder than any laws.

The outlook for other depleted species may be improving, or at least less dismal than seemed the case a little while ago. Take the roseate spoonbill, the most generally eye-catching of all the native wading birds of Florida. Somehow, the spoonbill has held on in southern Florida though for decades its numbers there were almost vanishingly small. I spent 30 years in the state thinking of spoonbills as virtually legendary, and when I finally saw my first one it was down in Honduras. Now, however, you can see spoonbills almost any day at low tide on the mud flats around Flamingo.

Some people believe that the flamingo is even more spectacular than the spoonbill, and I agree. The flamingo is probably the most spectacular bird in the world. Few sights in the bird world can equal the fields of pink flame that feeding flocks of flamingos make, or the incredible spectacle of the nesting colonies, with acres of rose-colored birds, each perched on a stumpy clay mound.

Unhappily, however, flamingos don't nest in Florida. The captive flock at the Hialeah Race Track near Miami has somehow been wheedled into breeding there, but there are no natural nesting colonies in the state—and perhaps there never were. Nobody can say why, but predation was probably the reason. In Florida a big conspicuous edible bird, with the strange habit of nesting on a short thick pedestal on a mud flat, has far more potential enemies than there are in the West Indies. The abundant raccoons of Florida could, alone, account for the ab-

An Everglades kite prepares to devour an apple snail, the only food it eats. The dwindling snail supply threatens this bird's extinction.

sence of wild flamingo breeding colonies. Some flocks used to come over from the Bahamas at molting time, when they were particularly vulnerable to predation.

Those that turn up here and there in Florida nowadays are thought to be birds that escaped from captivity—particularly from Hialeah, where they were established 40-odd years ago. A flock of 19 birds escaped from Hialeah in 1931—their custodian had neglected to clip their wings or pinion them—and smaller groups have subsequently been lost there and at other tourist attractions, though some unpinioned birds have been persuaded to remain. We will have to get along without this admirable bird as a true resident, though, and that is a real deprivation.

The story of the great white heron is more encouraging—if encouraging is the right word to describe the consolidation of as tiny a range as that of the great white. The total United States population of this tall, raucous-voiced bird is confined to southernmost Florida, from southern Biscayne Bay down the Keys, out into Florida Bay, and up around the west coast to Everglades City. According to Alexander Sprunt IV, research director of the National Audubon Society, the great white heron is "in fine shape." He is speaking comparatively. He estimates that there are probably about 2,500 to 2,800 birds in existence, and is pleased with the number because it used to be a whole lot smaller. One peculiar attribute of the great white heron that, in view of the dismal influx of people into its range on the Florida Keys, may be necessary equipment for survival is its willingness to fraternize with man. It often ventures into waterside human neighborhoods. Down in the lower Keys, where the Great White Heron National Wildlife Refuge has been created, the birds sometimes stroll about from backyard to backyard in the new subdivisions, accepting spare victuals from the residents. A few days ago I saw one standing beside a man on a dock, hopefully watching him clean a fish. The great white heron has come a long way since my first visit to Florida Bay in 1931, and it may live on.

The prospects of the reddish egret are less comforting. The first of these rare birds that I ever saw was, like my first spoonbill, not in Florida but on a shallow-water marl flat on Great Inagua Island, toward the southern end of the Bahamas. What I beheld there seemed to my casual and unexpert eye to be an ordinary heron of some kind, probably a young little blue. It was foraging in ankle-deep water in company with a sprinkling of mature little blues and a thin stand of roseate spoonbills. Suddenly what I assumed to be a young little blue began cavorting as I had never seen a heron cavort before, dashing around the shallows, abruptly stopping once in a while to stab at the water, and then gulping down some small creature. This behavior was without precedent in my heron lore, but there was a bird man with me and he said casually, "that's a reddish egret out there," and I vaguely recollected two things I had heard about this species: that they are not red but come in white and dark phases; and that their foraging behavior is peculiar.

Actually, the behavior goes far beyond the merely peculiar. In fact, watching the one on the Inagua flat that day it was hard for me to understand why the other normal-acting herons and spoonbills out there put up with such clowning. It bordered on the antisocial, or seemed to. Later, however, I read some papers on heron behavior by Andrew Meyerriecks, a zoologist at the University of South Florida, and learned that the reddish egret's eccentric feeding habits are probably an adaptive device that allows the species to avoid competition with other herons. According to Meyerriecks, if wading birds of different species all fed in the same way, they would spend more time threatening each other and squabbling over morsels of food than actually feeding.

Wading birds are not the only kinds of birds that are in trouble. Perhaps the most seriously threatened of all the Everglades birds is the Everglades kite, or snail kite. It qualifies as endangered by any definition. The range of the subspecies it belongs to is small—it is confined to southern Florida. Its total population is minuscule, probably not having been more than 150 birds in recent years. And its numbers are clearly declining. According to Paul Sykes of the U.S. Fish and Wildlife Service, the colony was down to probably no more than 20 to 30 birds in 1972, and these had been scattered by drought and fire into places far out of their usual range. There is no clear rescue plan in sight.

The chief cause of the snail kite's trouble is the decimation of Everglades snail populations by combined natural drought and human mismanagement of the water system. This species of kite is the most specialized bird of prey in North America. Its diet consists solely of a single species of fresh-water mollusk, the apple snail. The apple snail, which is about the size of a golf ball, is one of the group that wall themselves off in their shell by means of an operculum—a sort of door—common to many other snails. Unlike most operculate snails, however, the apple snail has a lung, and comes to the surface to breathe. This is the habit that brings it within reach of the snail kite. The bird pries off the operculum and with the long tip of the upper beak severs the muscle that holds the snail in its shell. Then it picks out and swallows the snail in chunks. The ranges of the two animals coincide closely. The survival of the kite population in Florida will depend on what can be quickly done to stabilize the apple-snail population.

Except for the masses of apple-snail eggs that conspicuously decorate the stems of emergent water plants, fresh-water mollusks go mostly unobserved in the Everglades. There is little evidence of the many-faceted ecologic role played by snails, mussels, pea-clams and the tiny limpets that huddle like wan barnacles against any smooth submerged surface. But the mollusks are down there in millions, filtering plankton, scraping algae, and industriously building themselves into food for larger creatures. Not many Everglades animals are strongly specialized eaters of mollusks, but various reptiles, amphibians, fishes and mammals eat snails, mussels or both, and if more were known about the flow of energy in the Everglades, snails would probably prove to be one of the most important links in the food chains there. Bullfrogs eat them regularly, and so do turtles, young alligators and many fishes.

Otters like snails too. At least our pet otter was avid about them. His idea of a pleasant outing was to be driven to a little pond that was full of coontail moss and ram's-horn snails. There he would porpoise about in the weed-choked water for an hour or more, crunching the snails so loudly we could hear him from where we used to sit in the pondside shade. Not long ago my wife and I saw a young wild otter in some kind of altercation with a limpkin. It was down on the Pinecrest Road, at the south edge of the Big Cypress, during a hard rain. From some distance away we noticed in the middle of the road two live, dark forms of about the same size but very different shapes. They appeared to be tilting at each other, and when we drove up close we saw that a half-grown otter and a limpkin were contending over some round black object. When I stopped the car and walked toward them, the limpkin flew off and the otter loped into the ditch beside the road. I walked over to the object of their strife. It was a big apple snail, still alive. The otter had not had a chance to take the snail in its mouth for crunching, nor had the limpkin gotten close enough to grab the snail with its beak. The snail, its operculum tightly closed, did not have a mark on it.

The feeding habits of the limpkin and two other mollusk-eaters, the Everglades kites and the boat-tailed grackles, have been the subject of an exhaustive study done by Noel and Helen Snyder. Their examination of this one section of a wetland feeding web shows how three very different species of animals can converge upon a food resource without harmful confrontation, and so illustrates an important principle of evo-

Two adult crocodiles bask on an Everglades mudbank. A narrower, pointed snout distinguishes them from the more numerous alligators.

lutionary ecology. The only open conflict the Snyders recorded among the three mollusk-eaters occurred between grackles and kites, or between grackles and limpkins. It usually was instigated by the grackles. Boat-tailed grackles—glossy, dark-plumed relatives of the blackbird—are omnivorous and versatile feeders. Through most of their extensive geographic range they are by no means dependent on or even especially partial to mollusks as food, but in parts of the Everglades apple snails predominate in their diet. They either catch the snails themselves or steal them from limpkins. Between limpkins and kites the Snyders observed no conflict, presumably because the two hunt in different ways. The kites cruise on the wing, flying at low levels over open, deep water. When they sight a snail near the surface of the water they drop fast and seize it in one extended foot. Limpkins avoid the open places, hunting usually in dense aquatic vegetation; because they move on foot they stay away from deep water. Wherever the Snyders found grackles feeding heavily on apple snails, the birds were hunting in shallow water like the limpkin. The grackles would search purposefully, sometimes leaning to lift a lily pad and inspect its undersurface.

Although the three birds' sharing of a particular food source in a particular area would appear on the surface to be ecological competition, it is not. To be competitors in a strict sense, they would have to feed in a way that decreased the common food supply, and this happens negligibly, if at all. Their predation causes only slight fluctuations in the snail population. Far greater effects on the numbers of snails are produced by the recurrent disasters of drought and fire, which kill the snails, drive limpkins out of their haunts, and in 1971 scattered the Everglades colony of snail kites as far north as Georgia.

Along with endangered birds, southern Florida and Everglades National Park serve as a refuge for other animals whose future seems uncertain. Two of these have become widely known test cases. One is the manatee or sea cow; the other is the American crocodile.

Although nearly all the early Florida explorers used to take home exciting tales of alligators, it was not till after the Civil War that the presence of crocodiles in the region became generally known. This seems strange, because the two are not really very much alike. The snout of the crocodile tapers much more sharply, it has the big ostentatious teeth in its jaws arranged in a different way, and in habits and temperament it is not at all like an alligator. Nevertheless it took a long time for Floridians to become aware there were crocodiles among them.

Once this was known, it took a lot less time for them to reduce the crocodile population to a tiny colony. In the early 1870s the range of the crocodile extended from Palm Beach on the Atlantic Coast to Key West, and up the Gulf Coast to Charlotte Harbor. Today nearly all of the few remaining crocodiles live in the northeastern corner of Florida Bay.

As with most of the other animals that are in danger of disappearing from Florida, the loss of the crocodile would not automatically mean the extinction of the species. It is pan-Caribbean in distribution and still survives in Cuba and Jamaica, along the mainland Caribbean coasts from Mexico into Central America. But little comfort can be taken from this wide range, because practically everywhere outside Everglades National Park crocodiles are under fearful pressure. The pressure comes from both hide hunters and the puerile "sportsmen" who get themselves guided out to a safe, shady place from which they can comfortably shoot a basking crocodile with a deer rifle.

From the time Everglades National Park was established in 1947 it was pretty clearly the only hope for saving the crocodile in the United States. For a while the hope was feeble, but now there seems some chance that it may be realized.

I recently went with park biologist John Ogden and ranger Stan Robbins to look at crocodile nests in the southeast section of the park. We saw five nests. Any clear expanse of shore appeared acceptable as a nesting ground. Some of the nests were in marl, some in friable shell sand. One had been dug down from beach level while the others had been excavated in mounds piled up by the crocodiles during several prenesting visits. One of these was 5 feet high and 20 feet across.

Since the recent spread of efforts to raise alligators on farms and ranches, quite a lot has been learned about the nest-building activities of the female alligator. Practically nothing has been known about the habits of the crocodile, however. In Central America the species makes its nests on seaside beaches or in the sand of river bars. A young friend of mine in Costa Rica got chased by a big female that built her nest at one end of a beach at Tortuguero. Until then, I had never known whether female crocodiles show the extraordinary parental concern for their nests and young that so distinguishes female alligators among reptiles. John Ogden's observations now show that they probably do.

Censuses that Ogden has taken suggest that the Florida crocodile colony has stabilized, and may possibly even be slowly growing. He estimates that there are probably 300 to 400 crocodiles now in the park, and that about 20 females nest there each year.

Submerged in salt water, a manatee, or sea cow, forages for aquatic plants. A one-ton adult eats about 200 pounds of vegetation a day.

The crocodile's position has no doubt been strengthened by the severe penalties for alligator hunting. Poaching is harder to get away with than it used to be. Certainly few people go out cold-bloodedly nowadays to hunt crocodiles for their hides. So the drain on the adult crocodile population comes mainly from inadvertent entanglement in the nets of mullet fishermen. Catching a crocodile in a mullet net is a fairly traumatic occurrence, for the fisherman as it is for the crocodile. Only a sawfish can make a worse mess of a net. In days gone by, the task of removing a croc from net webbing was performed with little or no sense of stewardship for the endangered species. But now, John Ogden tells me, some fishermen who used to kill the crocodiles they caught "to thin them out" actually cut them free.

The other chief cause of loss in the crocodile colony is nest robbery. Crocodiles nest in the marl or shell of beaches along the shores of eastern Florida Bay and upper Key Largo. Since most of the nests are mounded, they are easy to locate. Every year Ogden finds some of them destroyed by human predators, partly, he thinks, out of vandalism, but in some cases to get eggs from which to hatch little crocodiles for illegal sale. On the strength of this information the park management has recently closed three creeks in the prime nesting corner of the Bay shore to all but authorized persons. So the crocodile population of the park is probably the best protected colony of the species anywhere, and at the rate at which wilderness is being invaded in the tropics, it is a main hope for the survival of the species.

Another Florida animal whose future almost surely depends heavily on the protection it gets in the park—not as a year-round refuge, perhaps, but as a seasonal migration station—is the manatee or sea cow. There are parallels between its case and that of the crocodile. Though sea cows are more widespread than crocodiles in Florida, the total population in the state is probably not much bigger; and the two animals have somewhat similar distribution outside United States waters.

One could argue that the case of the manatee is the most important survival problem the Everglades National Park is involved in. I realize that it is risky to speak of degrees of importance where extinction is threatened. Obviously, all creatures are equal in the sight of the Lord. But I suggest that sea cows rank higher than most creatures as candidates for human solicitude. My criteria are their steady depletion during the last century; their present perilously low population level; the uniqueness of the order to which they belong; their enormous size

—they weigh up to 1,500 pounds—and their astonishing appearance. But all the reason anybody really needs for saving manatees is that they are fabulously outlandish and lovable beasts. Not every visitor to the Everglades National Park may have the privilege of seeing a sea cow; but everyone who does is bound to be glad there is such a place for such a beguiling creature.

Manatees belong to the Sirenia, an order of mammals of which the dugongs of the tropical Indo-Pacific area are the only other surviving members. Although they are derived from a common terrestrial ancestor, and are distantly related to elephants, manatees are strongly modified for aquatic life. A sea cow's body, while tremendous, is wholly streamlined. The front limbs are paddles, the back limbs have completely disappeared, and there is a broad, horizontally oriented tail flipper. The upper lip is a short, thick, snoutlike implement used in grazing on submarine vegetation. The face is whiskered and wrinkled.

There is a persistent tradition—reflected in its scientific name—that the manatee was the mermaid of the early mariners. If so, then one can only pity those chaps. Manatees are splendid, amiable animals, but they are ugly as sin, and one wonders how their one ladylike attribute—a pair of pectorally placed teats—could have so bemused the wistful old sailors as to evoke the mermaid legend.

The most encouraging development in the survival outlook of the Florida sea cow—aside from the establishment of Everglades National Park—has been the reappearance of a good-sized winter colony in the headwaters of Crystal River in Citrus County. According to a careful study of the herd there by Daniel S. Hartman, the sea cows turn up in the river regularly after the first cold spell of winter, apparently attracted by the warm spring water at the source of the river. Most of the adult manatees have numerous propeller scars on their backs. Dr. Hartman has made ingenious use of this phenomenon. By identifying the different scar patterns of 50 individual manatees, the way an ornithologist bands birds, he has been able to track them during two seasons and thereby study their movements and behavior.

This reappearance of manatees in the big springs of the peninsula seems almost incredible. I have known Crystal River well since the 1930s and never saw a sea cow there until 1967. No doubt the resurgence was produced partly by improved law enforcement against poachers and partly by a marked increase in submerged river vegetation. Suburban development around the headwater springs has grievously overenriched the river, and the burgeoning water plants seriously interfere with boat traffic. Manatees graze on the plants, and that makes them welcome visitors. Dr. Hartman's important data are now being used by Friends of the Earth as the basis for an effort to have the Citrus County springs declared a manatee sanctuary.

Another factor in the increase of manatees in the midpeninsular springs may be the slowly improving protection the sea cows are getting far to the south when they move on to Everglades National Park. The Crystal River manatees are there only in winter; when spring comes they disappear. Since this is when they most often are seen in southern Florida, the assumption is that a seasonal migration occurs. If this is a fact, then success in saving the Crystal River sea-cow herd will depend on protection in its more southerly summer range.

Fortunately, the most important part of that range appears to be within the limits of the Everglades National Park, in the labyrinthine mangrove country east and north of Whitewater Bay, and the Cape Sable area. Park pilot Ralph Miele sees them often there, and recently located a herd of 20 or more. They are often seen with calves there in spring and summer. They also are regularly reported by people traveling among the Ten Thousand Islands south of Everglades City. If these are the same manatees that turn up in November or December in Crystal River—and if the proposed Citrus County sanctuary becomes a reality—then the whole Gulf Coast manatee colony might be saved.

In 1885, in his book *Life and Adventures in South Florida,* Andrew Canova had a dire prediction for the Whitewater Bay manatees: "In days long gone by, the Seminoles living . . . near Cape Sable killed the manatee, jerked the flesh and sold it to the Spaniards at a good price, and ten years ago the meat could be bought at fifty cents a pound. Of course the animals are becoming far too scarce to admit of its being sold at all. There is no doubt that the manatee is fast becoming an extinct animal. . . . The sea cow will pass out of existence . . . and the only remaining trace of its former existence will be a few old bones."

Canova wrote that almost nine decades ago. The sea cows are not yet gone. It is clear that only great vigilance is going to save them, however, and that the main hope for achieving this is the calving ground in the mangrove bays and mazes of Everglades National Park.

An Audubon Sampler

Apart from the alligator, no creatures are more symbolic of the Everglades and their environs than the curious and beautiful birds that live there. And no one has done more to dramatize them—and bring them to the consciousness of the public—than the 19th Century painter-naturalist John James Audubon.

As in his wildlife studies in other regions of the United States, Audubon's superb visual record of the birds of the far south benefited from a method he devised to ensure true, lifelike representations. Because a bird's colors fade soon after death, Audubon used freshly killed models; by passing a sharpened flexible wire through the model, he was able to bend it into a pose characteristic of the living bird.

Produced in the 1820s and '30s, these paintings served as the basis for the engravings in Audubon's monumental work, *The Birds of America*. In time most of the original watercolors were acquired by the New-York Historical Society, which has exhibited them only intermittently to protect them from fading. It was from these originals that the photographs on the following pages were made. The studies of the anhinga *(page 104)* and the common egret *(page 107)* are here published for the first time anywhere; later paintings of the same birds were chosen for *The Birds of America*. In their freshness and vitality, all nine studies capture Audubon's intense feeling for nature—a feeling, he recorded in his journal, "bordering on phrenzy."

Audubon was deeply disturbed by the senseless slaughter of birds for the plume trade, which nearly resulted in the extinction of the common egret and the snowy egret, among other birds. The artist's revulsion at such wastage helped spur a demand for legislation, first passed in 1910, that prohibited the sale of plumage for ornamental use. But Audubon's unique contribution to the birds he loved remains his gift for tireless observation, expressed not only in his art but in his writings. Though not scientifically trained, he became a great ornithologist and a meticulous compiler of information, and was well aware of his contributions. "What a treat for me," he once wrote his publisher, "to disclose things unknown to all the world before me. . . . The truths and facts contained in my writings and in my figures of Birds will become more apparent to every student of nature."

Far more common in America today than its endangered cousin the brown pelican, the white pelican is one of South Florida's largest birds, with a wingspread up to nine feet. As Audubon painted it, the bird appears to be standing on a shoreline at dusk, holding its catch in its large pouch after a fishing expedition in offshore waters.

Amply endowed with soft, fluffy white feathers ideal for adorning hats and fans, the snowy egret was once a prime target for Florida hunters. Almost extinct 50 years ago, the bird has made an impressive comeback; at nesting time in spring, snowy egrets now safely congregate by the thousands.

The roseate spoonbill, one of the Everglades' most spectacular wading birds, is named for its spatulate bill, some six to seven inches long. Audubon noted "a considerable degree of elegance" in the way the bird uses its bill, swinging it from side to side as it munches insects or small shellfish.

The green heron abounds in Florida, but its protective coloring makes it hard to detect. As the painting shows, this shy bird, smallest of American herons, easily blends with the marshy vegetation growing in the shallows through which it wades in search of its staple diet of insects and small fish.

Another heron, the reddish egret, is never red, but most often gray, with a rusty brown head and neck; some are white (Audubon mistook this for the mark of an immature bird). But all reddish egrets have flesh-colored, black-tipped bills and a clownish walk; sometimes they lurch as if drunk.

A bird of diverse talents, the anhinga goes by three aliases: snakebird, for its serpentine neck; American darter, for its jerky movements; and water turkey, for the way its tail spreads in flight. It is as adept in the water as in the air, swimming either entirely under water or with head and neck showing.

The limpkin's name derives from its jerky gait as it prowls its marshy habitat hunting for the fresh-water snails that provide its main food. But the limpkin does not always limp. As Audubon noticed, its feet are so broad that it can walk easily in mud—and even on the leaves of aquatic plants.

About four feet tall, the regal great blue heron is one of the largest wadin birds in North America. It spends its entire life in Florida, building what amounts to a permanent home in its nesting areas near Everglades waterways; it will occupy the same bowl-shaped nest of sticks and grass year after year, renovating and addin to it with each passing season.

The mating-season plumage of th common egret—50 delicate long whi aigrettes that grow between i shoulder bones and extend beyond i tail—was nearly its undoing in the 19t Century heyday of feather-decorate millinery. Now multiplying in man parts of Florida, this most beautiful the herons has become a remarkab friendly, easily approached bir

5/ Rock, Snakes and Snails

The whole world of the pines and of the rocks hums and glistens and stings with life.

MARJORY STONEMAN DOUGLAS/ *THE EVERGLADES*

In 1838, during the Seminole War, a U.S. Army surgeon named Jacob Rhett Motte accompanied Colonel William S. Harney's Second Dragoons on an expedition to hunt down the elusive Chief Abiaka, also known as Sam Jones, as he retreated inland from Fort Dallas, the present site of Miami. Dr. Motte, a South Carolinian, was an articulate reporter, and his account of crossing the limestone pinelands of southern Florida enhances one's admiration for the Indians who traveled that country ahead of the U.S. Army—and for the Army too, in spite of the fool's errand they were on. It was an extraordinary and somewhat forbidding landscape, as Dr. Motte's impressions make plain:

"We pursued our way through a pine-barren, the ground being formed of coral-rocks jutting out in sharp points like oyster beds, which caused us great suffering by cutting through our boots and lacerating our feet at every step . . . as if we were walking over . . . a thick crop of sharply pointed knives. The whole of this part of Florida seemed to present this coral formation protruding through the surface of the earth, and which rendered it impracticable for horses and almost impracticable for men unless well shod. We were puzzled . . . how the moccasined Indians got over such a rough surface until we subsequently ascertained that they protected their feet . . . by moccasins of alligator hide when in this part of Florida. . . . It was certainly the most dreary and pandemonium-like region . . . where no grateful verdure quickened, and no generous plant took root—where the only herbage . . . was stinted and the shrubbery was bare, where the hot steaming atmosphere constantly quivered over the parched and cracked land —without shade—without water—it was intolerable—excruciating."

And with all that they never caught Sam Jones.

It is easy to understand why these flatwoods, visited in the bleakness of the dry season, should have distressed a homesick and poetic doctor accustomed to the magnolia and live-oak groves of Charleston. But a lot depends on one's point of view, and some people react differently to the rocky pine country. I used to go down there every chance I got, and though it was mainly reptiles I was after, other things there made memories too. One of these was just the look and sound of the pines, with their trunks a little more twisty and gnarled of limb than the pines I knew farther north, and with their sprays of thin needles singing in a special way against the particular blue of a southern Florida sky. The pines eke out sustenance and a water ration by spreading their roots widely over the honeycombed rock and invading every depression in it. To see what a plant is up against in this region you ought to look at a pine tree that has been pushed over by a bulldozer—at the upturned disk of desperate roots that in their sculpturing reflect every irregularity in the crazy stone surface they had grown on. Nevertheless, some of the pines used to reach heights of 80 or 90 feet.

These woods have undergone more widespread destruction by man than any other landscape in southern Florida. The pine timber, once a prime resource, has been repeatedly cut everywhere except in some of the Everglades National Park lands on Long Pine Key. Moreover, fire has been kept under better control in recent years and, in fire-free areas, broad-leaved hammock vegetation is invading the old domain of the pines. More relentless than either of those factors has been the spread of the cities of the lower east coast, which have engulfed most of the landscape. So the limestone pinelands are mostly gone now; and if the National Park had not preserved some sizable samples, in more or less mint condition, this fascinating landscape would only be read about or seen in a few old photographs, mostly bad.

Actually, there is no landscape like this one anywhere in the world except in the Bahamas. It is an open forest of slash pine growing on a limestone ridge known variously as the Atlantic Coastal Ridge, Rockland Ridge, Pineland Ridge or Rock Rim. This slightly elevated region is a series of outcrops of Miami oolite, a limestone formation—not an old coral reef, as Dr. Motte thought—that extends intermittently southward along the east coast for more than 50 miles, from north of Fort Lauderdale down to Florida City. There it curves westward into Everglades National Park and ends far out in the Glades on the pine island called Long Pine Key. The Rock Rim varies in elevation from 20 feet above sea level at the northern edges of Miami to less than two feet near its westernmost extreme in the park.

The surface of the ridge is thinly covered with sand or completely bare. Much of it is so fantastically eroded, pitted and perforated, and in the dry season seems so inhospitable, that one wonders how plants or animals could ever have found life ecologically feasible there. But the place is almost free of winter, and each dry season is sure to end with rain, because most of the ridge is located near the Gulf Stream downcurrent from tropical Cuba and in the same moist storm track as Cuba. Besides that, the ridge is not only the highest but also the oldest land in southern Florida, part of it having stood above the seas that covered the land around it for tens of thousands of years. There has therefore been a respectable span of time for any tropical plants that could get to the place, and could put up with its rigors, to establish themselves.

To botanists the limestone pinelands are in some ways the most exciting landscape in Florida. Though pines, palms and palmettos predominate, the ridge communities also include a great variety of small tropical trees and shrubs and herbaceous plants. The strong tropical contingent, which occurs all through the pine-wood understory, beneath the main canopy of the forest, is mixed with numerous representatives of the Temperate Zone flora.

Among these plants are the partridge pea, which has bright yellow flowers and leaves that fold up when touched—perhaps reacting to the pressure and heat of the hand's contact; and fire grass, which thrives in the aftermath of fire, with flower stalks eight feet tall. Another is the coontie, a kind of arrowroot with a bulbous root and bright orange-red fruit. The Indians gave it its name, and made flour out of the starch in its root. The plant, growing in plenty throughout the pinelands, was their staple source of carbohydrate food, and it was one of the factors that made the Seminole War the most frustrating the United States ever got into, until we went to Vietnam; the elusive Indians, pursued by white troops, might abandon their stores of coonties, but there was always plenty more to be gathered.

Of the tropical species in the pinelands, one of the most striking is the gumbo-limbo. In Honduras and Nicaragua the tree is called the naked Indian, because of the smooth, brightly copper-colored bark that peels away in thin curls like birch bark, only thinner. Other widespread tropical species of the pinelands are poisonwood, a relative of poison ivy with a similarly rash-raising juice, and blolly, a shrubby, scaly barked member of the plant family known as four o'clock, for their way of closing their flowers during the afternoon.

Whatever their origin, nearly all the plants of the pinelands show some kind of adjustment to fire. The slash pine itself, for example, germinates best in the mineral soil produced when a ground fire burns off the duff. Though vulnerable as a seedling, it quickly becomes insulated by heat-resistant bark, and tight rosettes of terminal needles protect its growing points. Some pinewoods plants, like the coontie, resist fire by storing much of their substance in bulbous stem bases or roots that will support quick regeneration when the fire has gone by. Most of the plants of the community are either clothed in fibrous, nonflammable bark or leaf bases, or, like the grasses and pines, protect their buds with tight clusters of twig-tip leaves that burn partly away while the bud remains unhurt.

The other principal vegetation of the limestone ridge is the tropical hammock, an elevated isolated forest that occurs only where fire has been excluded. Like the hammocks of northern and central Florida, these tropical hammocks are composed mainly of broad-leaved trees, but in other ways they are very different. The southern Florida hammocks are usually somewhat lower, and botanically they are usually more diverse. The most striking difference is that in southern Florida trees in the hammocks are mainly of West Indian origin.

These West Indian trees and shrubs are the most conspicuous tropical element in southern Florida. Some hammocks are exotic-looking stands of almost wholly Antillean trees with marvelous names like strangler fig, pigeon plum, Madeira, bustic, torchwood, fiddlewood, nakedwood and paradise tree. In other hammocks, various combinations of these mix with live oaks, red bay, mulberry, hackberry or any of a dozen members of the hammock flora of northern Florida.

Like the pinewoods, the tropical hammocks of the Rock Rim show the tragic effects of man's spread. In Dade County alone, according to an estimate by John K. Small, there were once 500 separate hammocks. A single hammock five miles long and half a mile wide used to run along the shore of Biscayne Bay where Miami now stands, occupying the highest section of the Rock Rim. Most of it became real estate years ago, but a fragment of it is preserved more or less intact in Simpson

Park. This is well worth walking in, though it almost makes old naturalists weep to do so.

With the widespread destruction of the hammocks, the scientific —and esthetic—interest in those that remain is more intense than ever. Some fundamental questions about them remain to be solved. How, for example, are the sparse, limestone pinewoods replaced by a hammock forest with a closed canopy and a moist, dimly lighted interior? And what maintains hammocks once they are established? At least part of the answer to the second question seems to be that they shade out pine seedlings, and that the higher humidity in normal years helps to keep out fire, which can quickly destroy hammock vegetation. The more difficult problem is to explain how hammocks begin in the first place—how the shade and interior moisture they require are achieved before fire bats down the incipient hammock growth. Most botanists agree that the live oak is often involved in the earliest stages of the development of a hammock. Charles Torrey Simpson suggested that some of the Rock Rim hammocks probably began with a single live oak tree. Lodging as an acorn derived from who knows where, the tree sprouted, managed to get through several years without succumbing to the ground fires that swept the surrounding pinewoods, and gradually covered the ground with its own hard little fire-resistant leaves that in time provided moisture-hoarding shade.

What then keeps the fire away from the pioneer oak when it is very small is not easy to say. Often the site is the edge of a pothole or sink where water has already humidified the surroundings; or in some cases the rock surface of a site is so bare and so jumbled that ground fire dies without entering it. In any case, the oak gradually creates a little zone of fire resistance, and other hardwoods soon begin to exploit this. The berries and small fruits of increasing numbers of shade-tolerant, fire-tender tropical species are, over the passing decades, fortuitously blown or carried there by birds. Finally a diverse, clean-floored forest is formed, and this creates its own moist internal climate, and holds out all fires except the driving holocausts that come in the most drastically drought-ridden years.

As time passes, the tropical composition of the woods continues to increase. The tropical species that arrive, though vulnerable to fire, are all of kinds accustomed to competitive life in tight stands. Their roots outfight the roots of the pioneers and the deepening shade kills the seedlings of the first comers. Dr. Simpson noticed that a live-oak seedling is

Slash pines—which are locally known as Dade County pines—stand among bushy neighbors of saw palmetto. Burn resistant because of their moist bark, the pines and palmettos are sturdy and thriving despite charring by a ground fire. Occasional and limited fire actually helps the pinelands, killing the seedlings of competing hardwoods that move in from nearby hammocks.

seldom found in one of the more diverse tropical hammocks, even though old live oaks may persist there. He also pointed out that the old trees themselves become the special prey of the strangler fig. In any old hammock in which live oaks persist, many of them can be seen, as Simpson says, "enfolded in the stifling embrace of this terrible *Ficus*."

Next to the plants, the most conspicuously tropical living things in southern Florida are butterflies. No more convincing evidence of the nearness of the tropics is needed than the beautiful zebra butterfly, a member of a group as characteristic of the American tropics as spider monkeys and tapirs are. The larva of the zebra feeds only on the leaves of the passionflower, a woody vine with a showy flower of which parts were once thought by the Spanish to resemble aspects of Christ's Crucifixion. Though the zebra is not confined to southernmost Florida, it is most abundant there, and seems clearly at home drifting about the quiet hammocks under the naked-Indian trees on its long thin wings of gold and velvet black.

Hammocks are the best butterfly habitat in southern Florida, but you can see more of them in the open vistas of the pinelands. On any clear warm morning the viceroys, swallowtails, hairstreaks, blues, sulphurs, and a bewildering host of other species spread out over the saw palmettos beneath the pines, floating or dancing about according to their kind, hovering over wild crotons or tarflowers and gathering in mixed bouquets at the blazing butterfly weeds.

A few southern Florida butterflies have almost died out during the past few decades. The most famous case is Schaus's swallowtail, a drab brownish species that is now one of the rarest of North American butterflies. The beautiful atala, marked with brilliant iridescent blue-green, has also been reduced almost to extinction. Between 1925 and 1927 it disappeared from view; then, in 1959 a small new colony was discovered in the pinelands and by transplanting larvae from that population, an effort has been made to establish the species in Everglades National Park.

The tropical aura supplied by the butterflies was but one of the attractions of the Rock Rim country for me; another feature that particularly appealed to me, as a north Florida zoologist with a predilection for reptiles, was the presence of snakes slightly different from those I knew at home. For example, the snake known in the north as the blue racer is slate gray in the south; there, the yellow rings of the northern Florida king snake break up into a pattern of yellow specks, one on each scale; in tropical Florida a marvelous confusion of pinkish rat snakes replaces the red-blotched corn snake and brown-striped rat snake of the north.

As everywhere, collecting reptiles in the limestone flatwoods involves one in an orgy of turning over logs. In fact, other than random search, rolling logs is about the only known snake-catching technique. I have turned over a great many southern Florida logs in my time, and have found quite a few snakes that way. My liveliest memories of those ventures, however, are not of snakes at all, but of the array of scorpions and centipedes that live under Dade County logs.

Of scorpions there seem to be three kinds that a nonspecialist comes upon. One of these, whimsically called the slender scorpion, is a broad, redoubtable beast as long as your finger. To turn over a log and come upon a big female of this species, almost completely hidden in a swarm of newly hatched young, is a stirring experience; but not so stirring as stumbling upon a certain species of Dade County centipede. It is six inches long and a half inch wide, and has an astounding redundance of active legs. This particular species of centipede is to me the most unsettling animal in Florida, one of the very few creatures whose mere appearance completely undermines my professional zoologist's objectivity. I attribute this to having been crawled on by one as a child when camping in West Texas. Another creature you find under the Rock Rim logs is a kind of millipede, as long as the centipede and with even more legs, but a lot less demoralizing—for me, at any rate.

There are more engaging, though not necessarily more interesting, animals in the limestone pinelands of Everglades National Park, and some are prime examples of the versatility of the region as a refuge for wide-ranging species. The most hopeful place to look for a panther south of the Big Cypress Swamp, for instance, is said to be the edges of the pinelands bordering the Glades at the southwestern end of Long Pine Key. It is in these woods, also, that extraordinary aggregations of young bald eagles occur. According to park biologist William Robertson, up to 50 eagles or more, mostly two or three years old and not yet breeding, assemble to roost in the tallest trees of the park pineland.

Few of the mammals that turn up in tropical hammocks are indigenous; most of them are species that also occur in the rest of peninsular Florida. One of the inconspicuous mammal inhabitants of the tropical hammocks is the wood rat, a close relative of the pack rat of the western states. Not long ago, in the dense low hammock on upper Key Largo, I came upon a wood rat's nest—or lodge—that looked just like the work of a small, somewhat inept beaver. It was a stack of twigs and billets piled waist-high against a gumbo-limbo tree.

The reptiles and amphibians of the hammock are also largely representative of the fauna of the Florida peninsula as a whole, although the tropics are represented by two little geckos and by a couple of lizards of the genus *Anolis,* called chameleons because of their ability to change color. True chameleons belong to a different family. Hammock reptiles, like most deep-woods animals, stay mainly out of sight. Still, you may come upon a surreptitious blue-tailed skink, and if you look closely you may see a tiny ground skink scrambling away in the leaf mold. With luck you will find a graceful rough green snake, a rat snake, a yellow-speckled kingsnake, or a poisonous, gleaming banded coral snake. Out near the hammock edge, you will more than likely come upon one of the gray racers peculiar to the region. Back in the days when hammocks were more widespread and people fewer, you were likely to come face to face with the indigo snake. This big, shining blue and personable animal reaches lengths of nine or 10 feet, and vies with the coachwhip, the horn snake, and possibly the diamondback rattlesnake as Florida's longest serpent. And speaking of rattlesnakes, if you leave a hammock trail, it is well to walk with extreme care, because drought, fire and famine in the pinewoods, or flood in the sawgrass, often bring diamondbacks into the hammocks to forage for rats, deer mice or swamp rabbits.

Exquisitely varied in their banded color patterns, six tree snails of the species Liguus fasciatus browse on microscopic fungi and algae that grow on the bark of the hardwood trees in their homes in the Everglades hammocks. The two-inch-long snails owe their great range of coloration to longtime isolation. The treeless wetlands around the hammocks restrict the tree-snail populations to their own native habitats; after innumerable generations of inbreeding, the snails in some hammocks have evolved into local sub-subspecies. Fully 52 color variants have been found, and others may exist in unexplored hammocks.

The list of tropical animals in the hammocks will very likely lengthen when entomologists have done more field work. Nearly every entomologist who collects in Everglades National Park finds new species of insects or specimens of hitherto unrecorded West Indian kinds. That strange insects remain hidden in the hammocks is suggested by the elusive character of the Margarodes, or "ground pearls," that have been found there two or three times. These glistening objects resemble pearls of irregular shapes but all about the same size; they have been discovered buried in tiny caches in rock crevices or around the bases of hammock trees. Nacreous little lumps of natural jewelry, they are actually the waxy shells formed by a kind of scale insect; in some West Indian islands, where they occur more regularly, they are strung as beads. Ground pearls evoke the wonder and admiration of all who see them.

It is the same with tree snails. If there is a single species in which the essence of tropical Florida seems to be packed, it is that of the lovely, painted tree snails of the genus *Liguus*—the ligs, as they are known to

collectors. And they are as baffling as they are alluring. The ligs have excited—and confused—scientists for over half a century.

The Florida tree snails live almost secretly, and only in hammocks; their appeal to zoologists is based on the explosive microevolution they have undergone in these isolated tracts of hammock habitat. They are hardy, and able to withstand long periods of drought by sealing the openings of their shells to any smooth surface. But they feed, breed and flourish only where there is a steady supply of the microscopic fungi and algae that they scrape from the smooth trunks and limbs of such trees as Jamaica dogwood and wild tamarind. Good grazing occurs only in the moist shade of the tropical hammocks, and it is for this reason that these woods are the sole habitat of Florida *Liguus*. A lig in a hammock is often quite as isolated from contact with outside relatives as if it lived on an island in the sea. The snails' evolutionary response to this situation has produced a microzoogeographic classic.

Tree snails of the genus *Liguus* are known only in Cuba, Hispaniola and Florida. Those in Florida are obviously derived from Cuban ancestors. None is the exact equivalent of any Cuban form, however. In their classification of Florida *Liguus*, William Clench and G. B. Fairchild recognized four subspecies of a single species, *Liguus fasciatus*. It seems likely that each of them represents the separate landing of a Cuban snail somewhere in southern Florida. It is easy to see how the snails could be transported by either hurricane-blown debris or ocean currents, because once they have stuck to a limb and sealed themselves shut they will stay alive, though dormant, for months. Besides the four subspecies that reflect separate arrivals from Cuba, there are more than 50 named color variants in the southern Florida hammocks, many of them also showing slight differences in the texture or shape of the shell. Nearly every snail-bearing hammock has its own form or its own combinations of forms of lig. While all the 50 color variants occur as pure separate colonies, they are also subject to all sorts of hybridization and merging.

Even if the tree snails were merely animals in black and white they would excite biologists. But they are much more than that. Naturally ensconced in their habitat, they are among the most exquisitely beautiful of creatures. They are about two inches long and are colored in nearly any way that you might imagine. Some are clear white, or white with a pink spire, or with thin green lines around the spiral of the shell. Others are marked with combinations of yellow, orange, green, blue or various shades of brown. Some look like objects in porcelain; some seem carved out of the most highly variegated tortoise shell. Laid out in rows in a tray, tree snails are spectacular enough. To find one hanging in the forest twilight from the smooth limb of a wild tamarind tree is an unforgettable event.

The tree snails long ago stimulated the development of a cult—a polyglot, fanatic set of people who, though utterly diverse in background, shared a language most of us couldn't understand. They had a minute knowledge of the southern Florida terrain that, in the days before the advent of helicopters and Glades buggies, was truly extraordinary. Membership in this loose but rabid fraternity ranged from eminent scientists to commercial shell tradesmen and out-and-out snail hogs. No matter what the angle of their approach, these people were all similarly inflamed by the gemlike beauty of the snails, and shared the same fervor for hunting out secret virgin hammocks.

If the numbers of these lig collectors had grown at the rate at which Miami grew, there would be not one painted snail left hanging like a gem from a single dogwood limb in all of Broward, Dade, Monroe and Collier counties—except in the Everglades National Park. Everywhere outside the park the hammocks are disappearing, and with every lost hammock a colony of tree snails is destroyed. If people were turned loose in the few remaining hammocks to Easter-egg-hunt the tree snails as people gather shells on beaches, the rate of extinction would immeasurably swiften.

But a lot of snails are being saved in the park; and recently its personnel have embarked on a project to establish pure strains of all the threatened forms in hammocks within the park boundaries that now have no tree snails, and so to save these strains from extinction. Some people may feel uneasy over the introduction of outside snails into the park, on the grounds that any man-made change in the natural distribution of animals is deplorable. But in this case, with the creature involved so lovely and so clearly innocuous as *Liguus* is, and as sure to stay each in its own assigned little woods—in such a special case, with each planting purposeful, meticulously recorded and monitored, why not let the park go ahead and save the snails any way it can?

NATURE WALK / A Visit to Paradise Key

PHOTOGRAPHS BY ROBERT WALCH

From a distance Paradise Key is a dark green wall of tropical hardwood trees rising sharply from the flatness of the surrounding sawgrass plain. In Florida parlance Paradise Key is a "hammock" (a word possibly a variant of "hummock"), but in the same vernacular it is also called a key. Isolated amid a great sea of grass, it is no less an island than those more famous keys that curve out across the turquoise waters off the Florida mainland 20 miles to the southeast.

This tree island got its name back around 1900, when it was known only as a fabled place of unusual size and beauty somewhere out there in the uncharted Glades southwest of Homestead. Since then Paradise has been pinpointed. It is one of the biggest hammocks in the Glades, measuring a mile by a mile and a half, and it lies along the western edge of Taylor Slough (pronounced "slew") near the main headquarters of Everglades National Park at Homestead. Since the watery expanse of the slough provides a natural firebreak, the hammock has been relatively free from fire damage over the years—at least at its southern end, where it has also largely escaped human damage. The vegetation there is as close to being unspoiled as that in any tree island in the Glades.

Paradise Key is an ideal setting in which to absorb the peculiar enchantment of a tropical hardwood hammock. It is a cool, dry-footed oasis in a steaming wet landscape, a quiet woodsy place where the limitless scale of the Glades becomes comfortably man-sized. Though it is much frequented, with a Park Service building and even some paved walkways, there are parts in its interior where almost no one ever goes, and where no man has left any trace of his presence.

On a warm June day a faint rank of cumulus off to the south threatened a rainy afternoon to come; but for the moment the midday sun shone serenely on the approach to the key, glinting off the white trunks of a few twisted buttonwoods sticking up here and there above the sawgrass. Scrubby encroachers like the red mangrove, the buttonwoods' presence was a sure sign that something—perhaps the gradual lowering of the Glades' mean water level —had disturbed the equilibrium of the moist sawgrass community. Given time, the intrusive buttonwoods might even take over the wetlands here, but for now the sawgrass was

SAWGRASS, BUTTONWOODS AND (BACKGROUND) PARADISE KEY

in charge, and alive with the buzzing and chirring of countless insects.

Close inspection of the sawgrass revealed that some of the plants were in early bloom. The sawgrass near the key was not showy: the teeth that give it its name were much easier to feel than to discern, and its reddish-brown blossoms, subdued in this season, went almost unnoticed. But the few scattered flowers provided a subtle counterpoint to the sunlit green of the marsh.

SAWGRASS IN BLOOM

The sawgrass is not part of the hammock, but merely defines its limits. It was a relief to leave the glare of the road and plunge into the cool, dim greenness of Paradise Key's wooded interior. The first impression was of an overwhelming variety

DUCKWEED AFLOAT IN A SOLUTION HOLE

of vegetation—trees, shrubs, ferns, bushes everywhere seemed to crowd one another for growing room. The first stop along the path, however, was at a break in the vegetation—a large shady pond called a solution hole or a sinkhole—that offered a telling look at the hammock's eroding rocky underpinning.

A Porous Foundation

Cropping out around the edges of the pond was the jagged, pitted limestone called Miami oolite, the hammock's foundation, seldom more than two or three feet below the organic peat that serves Paradise Key as soil. The sinkhole, about 25 feet wide, is carved into this bedrock. It has been formed over the ages by rain water, charged with carbon dioxide and carbonic acid, that drops off the leaves and trickles down through the porous limestone, dissolving it and carrying it away in solution. The resulting hole, some 20 feet deep, is filled with clear, coffee-colored water that silently rises and falls with the changing water level of the Glades outside.

There was an air of magic about this lush, cool grotto and somehow the dappled light conjured visions of elves and fairies. Most elfin of all was the tiny duckweed that decorated the pond surface with a yellow-green appliqué design. Duckweed is one of the world's smallest flowering plants, and it floats unattached to the pond bottom, buoyed by leaves that measure no more than a quarter of an inch in diameter.

Spreading its branches across the water was a large pond-apple tree, a

AN UNRIPE CUSTARD APPLE

A SHY BOX TURTLE

A LUSH GROWTH OF FERNS

tropical species that would have nothing to do with the comparatively dry hammock were it not for the moisture provided by the sinkhole. The tree is also known as a custard apple, perhaps because its fruit, when ripe, has the consistency of blancmange. It is considered a delicacy by turtles. Now in late spring, though, the fruit was bright green and hard as any other unripe apple; one hanging from the leafy roof of this little Eden was comely, but untempting to the palate.

A rustle in the damp shade near the pond announced a common box turtle cautiously exploring for lunch. At the approach of human danger it pulled its head inside its shell and waited patiently. Despite the elegance of its black and yellow shell, which looked as if it had just been lacquered, the turtle was a prosaic note in this exotic place. It would not have seemed out of place poking around a maple woods in Illinois or a birch grove in Vermont, as some of its cousins do.

Many of the ferns around the pond also would look familiar in northern woods, but here they are bigger and much more prolific. The dark, narrow sword fern and the broader shield fern are the two most numerous plants in the hammock. They thrive wherever the ground is moist, in shade or brilliant sunlight, and are the leitmotiv of Paradise Key.

From the pond, the way led toward the key's interior, where the path was uncertain and the light even more subdued, but still capable of magical effects: the bark of a small gumbo-limbo tree had peeled away from the slender trunk and, backlit by a vagrant sunbeam, was glowing like a firebrand. The show lasted only a few moments—the sun moved and the fire went out. But even without dramatic lighting the gumbo-limbo, with its variegated pa-

SUNSTRUCK GUMBO-LIMBO BARK

pery bark, is one of the most distinct and recognizable trees in the hammock. It is a native of the tropics, and the origin of its name is obscure. Some people believe it comes from the Spanish-Dutch combination *gom elemiboom,* signifying "gum-resin tree," and was bestowed on the tree by West Indian colonists who recognized its major significance to the islands' natives: they collected its ar-

A STRANGLER FIG AND ITS FALLEN HOST

omatic resin to meet many of their needs, including solvents, medical ointments and even incense.

Beneath the peeling bark—the mark of an immature tree—the wood is soft and workable, and that is why many a gumbo-limbo ended up, back in more ingenuous times, spinning in circles as a merry-go-round horse.

A Bizarre Killer

There was nothing so whimsical about the strangler-fig tree that suddenly loomed in the path, its roots wound like a tangle of boa constrictors around a huge old live-oak trunk, which, toppled and dead, now belied its name. Certainly the fig is a bizarre plant. Though a true tree, it often spends years of its life growing like a vine until its roots spread down the trunk of its host, and finally dig into the ground. Though not a true parasite—taking neither nourishment nor moisture from its host—it nevertheless dooms the tree it grows on to inevitable slow death by constriction.

This specimen, however, had apparently not killed the oak it was entwining, but was simply using it as a prop—the oak had probably been felled years ago by a hurricane. The oak was the biggest tree encountered yet, and even dead was an impressive representative of one of the hammock's three foremost tree species (gumbo-limbo and strangler fig are the other two). Unlike them, the live oak hails from the temperate zone, no different from those serene old giants that grace the lawns and driveways of plantation houses all through the South.

BUTTONWOOD

WAX MYRTLE

WILD COFFEE

SATINLEAF

Scrambling over the rough-barked old hulk was no problem, but beyond it the going got slower. Now the undergrowth was much more dense, a bewildering disorder of different kinds of plants presenting an even more bewildering similarity of appearance. All the shrubs and young trees that grew here seemed to have smallish, dark green leaves with pointed tips and a smooth, waxy surface. This may not be a coincidence: many botanists believe their tipped design is an adaptation that allows heavy tropical rainfall to flow easily off the leaves. Such a leaf structure is useful for the plants but confusing to an untrained eye trying to tell one species from another.

A more careful inspection, however, revealed distinctions among the array of similar-looking plants. Within the compass of one glance was a scrub buttonwood bush, like the ones in the marsh outside, sporting little fruit like miniature hand grenades; wax myrtle, also called southern bayberry, and cousin to the northern variety, whose waxy berries make sweet-smelling candles; the vivid, deeply veined leaf of wild coffee, related to the domestic plant, but of no use in the kitchen; the satinleaf, which has a dark, oval leaf with a delicate russet underside; and various other plants whose names sound an exotic botanical litany—poisonwood, pigeon plum, lancewood, Jamaica dogwood, paradise tree, medicine vine, bullbrier, tetrazygia and dozens more.

Suddenly the dense tangles of undergrowth opened out into a place

A MATURE TROPICAL HARDWOOD FOREST

where there was very little brush at all: a mature forest of lofty trees, a shadowed harmony of muted brown and yellow-green.

As in so many other places in the Everglades, the reason for the abrupt change of scene is fire. In 1945, a year of great drought in the Glades, even Taylor Slough went dry, and wind-swept flames raged across it to engulf the northern part of Paradise Key. In the fire's wake a new growth of dense vegetation took over. The fire burned itself out, however, before it could ravage the southerly part of the key, and here the old growth of trees, tall and straight, has spread a leafy canopy aloft that largely keeps out the sun and eliminates all but the most shade-tolerant plants. Among them are the ubiquitous ferns and a few coonties—a low shrubby vegetable with a tuberous starchy root that was once a favorite food of the Indians.

In this mature forest the aggressive growth that pervades other parts of the key gave way to a sense of stillness and peace. Dead leaves provided a comfortable cushion. One of the six large live oaks that dominate the place made a fine backrest. The minutes passed serenely in that lovely place, and the steamy, teeming Glades, which edge up to the key less than half a mile away, seemed part of another world.

But there was more to see and do,

A GOLDEN ORB WEAVER AND HER PREY

A LUBBER GRASSHOPPER

CORN-LEAVED TRIPSACUM NEAR A LICHEN-SPLOTCHED TREE

and the way led north and west, toward the rim of the hammock and another change of scene. Now, no longer screened out by the forest canopy, sunlight flooded in and exposed some of the creatures of the key. At eye level, a large, marvelously constructed spider web fractured the sunlight with exquisite symmetry. It was the work of a female golden orb weaver, whose silk is stronger and finer than a silkworm's and whose instinct for design is spectacular. This spider's web was at least three feet across and she sat dead center, suspended five feet off the ground, waiting. A sudden twitch of the web galvanized her into swift attack on a hapless beetle that had blundered into the trap; just as quickly she was back at her place again, the beetle clasped in her jaws.

A Defensive Reaction

Above her a huge, armor-plated lubber grasshopper crept along a tetrazygia branch, munching whatever leaves it encountered—and then, when it was picked up, spitting them all out again in a defensive brown stream of fragments.

A cardinal flashed brilliant red and a faint breeze stirred the slender leaves of several tripsacum plants (which at some point in evolutionary history had shared an ancestor with golden bantam corn). Behind them stood a tree trunk that appeared splashed with white paint —which turned out to be a splotch of lichen decorating a lisoloma tree. Lichens of all textures and colors embellish tree trunks everywhere in the hammock, making many types of trees unrecognizable. The air plants create the same baffling effect.

No hardwood hammock is without air plants—also called epiphytes —and in some hammocks they grow in more abundance than anywhere else in southern Florida except Big Cypress Swamp. They have no use for roots in the ground, getting all they need for life from sunlight, air and rain. They perch everywhere, high and low, on trees, stumps, rocks, roots and twigs. In this part of Paradise Key the air plants' favorite hosts are the live oaks, whose rough bark gives them a good foothold. Sometimes they completely reclothe the tree trunks in green.

The most abundant air plants in southern Florida are the bromeliads, whose most conspicuous species, the

A YOUNG GIANT WILD-PINE AND RESURRECTION FERN GROWING ON A LIVE OAK

A STIFF-LEAVED WILD-PINE

stiff-leaved wild-pine, looks enough like its relative the pineapple to account for its similar name. Another cousin, the giant wild-pine, is the largest of the bromeliads, sometimes achieving a height of almost six feet. The resurrection fern, also an epiphyte and a member of a more ancient family of plants, curls up and shows only its brown undersides in dry times, then uncurls and appears bright green after a life-giving rain.

All these, plus an assortment of other epiphytes, including a variety of orchids not yet in bloom, were seen garnishing a single big live oak; they managed to give the tree the look of an overdressed dowager draped in her bright costume jewelry and green feather boas.

The most vivid aspect of this part of the key was the greenness of everything, whether high in the oaks, or at ankle level, where a saw palmetto shone like an emerald sunburst. Beneath its leaves, each of which spread out evenly from a central stem, was a tiny green rain frog, crouching on a leaf in ceramic immobility. It could easily have been a porcelain figurine but for the almost imperceptible pulse that beat in its throat. One of several kinds of frogs with adhesive toe pads that enable the animals to plaster themselves to Florida windows, it gets its name from the fact that its regular, throaty song is often heard after a summer rain shower.

A GREEN DARNER DRAGONFLY

A RAIN FROG

SAW-PALMETTO FANS

A Swarm of Mosquitoes

From the look of the sky at that moment, it appeared that the rain frog would soon have something to sing about. The softly piled cumulus that had seemed so far away earlier in the day was now poised directly over Paradise Key. Solid shelter was more than a mile away, and a wetting seemed inevitable, but there was a more disconcerting and immediate problem. Perhaps it was the sudden atmospheric change announcing the storm that brought out a merciless swarm of mosquitoes from hiding. The mosquitoes were in their prime at this wet time late in the spring season, but fortunately some dragonflies were in evidence, avidly doing their best to keep the mosquito population down. The most prevalent dragonfly was one known locally as a green darner, with a twig-shaped body two to three inches long and two pairs of transparent wings.

Their name derives from the darning motion made by the female's body when she lays eggs, and by the male's during copulation; but neither sex seemed interested at that moment in anything except eating, for they were purposefully darting and whizzing about in search of food. One finally settled down on a thin branch and immediately became nearly invisible; its body blended almost perfectly with the background and only its delicate fairy wings caught and shaped the sunlight.

After a minute of resting or digesting, the darner buzzed off again. A few minutes later the sun went too, and the rain came down.

It made no difference, really. A big gumbo-limbo provided shelter of sorts, and it was a treat to watch the bark turn dark red, like dyed tissue paper, as it got wet. All the small, tipped leaves shed water efficiently and indiscriminately; within minutes everything in the hammock had been thoroughly and satisfyingly soaked.

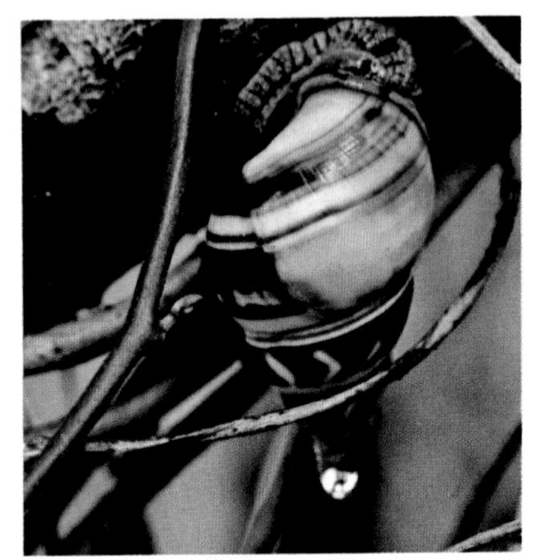
A RAIN-BEJEWELED TREE SNAIL

The rain was a good one, hard and noisy, with a few thundercracks to give it authority. This is what late spring in the Glades is all about: rain to make things grow, rain to fill up the marshes so the animals can spread out across the Glades to start a new life cycle. Rain is life itself in the Glades, and no creature there minds getting a little wet.

The shower lasted about 15 minutes and then, in the way of tropical rains, stopped suddenly, leaving the woods dripping and the ground soggy. The noise of the rainstorm yielded to the typical quiet of Paradise Key, in which the voices of the hammock's creatures were increasingly audible. Most distinctive of all was the slow, deliberate *ack-ack-ack* of rain frogs celebrating the wetting. Mingled with their song in sudden cacophony were the harsh scream of a great blue heron, the cardinal's persistent *what-cheer-cheer-cheer,* and occasionally the percussive *tonk-a-tonk-a-tonk* of a pileated woodpecker attacking a dead tree.

The Spectacle of a Snail

It proved to be a lucky rain, becase no walk in a tropical hardwood hammock is complete without a tree snail, and now here, just above eye level, was a lovely specimen with brown and white and yellow stripes. These highly polished bands

A LITTLE BLUE HERON ON THE HUNT

WATER-SHIELD PADS AFTER A RAIN

of color, following the whorled contours of the mollusks' shells, are the marks that identify each snail's variety—in this case *Liguus fasciatus walkeri*. Many another tree island, smaller and more isolated than Paradise Key, supports only a single and unique variety of tree snail and holds the strain to a color evolution all its own. But *walkeri* is only one of the varieties found in such a large hammock as Paradise Key, and a common one for all that. Yet, whatever the snail lacked in exclusiveness, it more than made up for in the brilliance of its shell, enhanced by a molten crystal drop the rain had left gleaming at its tip.

When interrupted by the rain the snail had been working its way along a tree branch, browsing on the fungus and algae growing there, leaving a tiny furrow behind it. Now it moved again, the raindrop fell and the snail resumed its grazing.

A few squishy steps led to the end of the path and the edge of the open Glades. A low bank marked the rim of the hammock. Beyond it, pads of water-shield plants cupped the rain in their water-repellent leaves like pools of mercury. Out in the sawgrass a little blue heron posed elegantly and solemnly, its piercing eye peeled for a fish dinner. And beyond that the great marsh stretched out to the low green line of another hardwood hammock in the distance. The rain cloud rolled on to the north, leaving the water level in the Glades a tiny but vital bit higher. Behind, from a fig tree in Paradise Key, a vireo sang its evening song.

THE VIEW FROM THE EDGE OF THE HAMMOCK

6/ The Tropical Borderland

Green, grateful mangroves where the sand-beach shines—/Long lissome coast that in and outward swerves. SIDNEY LANIER/ *A FLORIDA SUNDAY*

One sign that the tip of the Florida peninsula is edging down close to the tropics is its coastal fringe of mangrove swamp. The Florida mangroves are among the most luxuriant in the world. Mangrove forest straggles north to Cedar Key on the Gulf Coast and to St. Augustine on the Atlantic side of Florida, but the farther south you go the finer the trees look, and along the southernmost edges of the mainland the mangroves are the equal in density of any to be found anywhere.

The word mangrove is apparently a contraction of *manglegrove*, derived from the Spanish name for the tree, *mangle*, combined with the English *grove*. Strictly speaking, the name mangrove refers to the 60-odd species of maritime shrubs and trees that belong to a single botanical family, *Rhizophoraceae;* the red mangrove (*Rhizophora mangle*) is the only Florida species. But the term is more loosely given to other, unrelated trees that grow in tidal habitat in the tropics, and the name also applies collectively to the various types of swamp and forest vegetation that these trees build. A common way of life at the edges of the land has molded them into similar habits and structure.

Although *Rhizophora* is the only true mangrove in Florida, the mangrove forest there includes three other salt-tolerant shore zone trees, two of which also bear the name mangrove. One is *Laguncularia racemosa*, the white mangrove; the other is the black mangrove *Avicennia germinans*, which produces the strange-looking breather roots known as pneumatophores that stick up out of the mud beneath the trees like a multitude of quills or little cypress knees. White mangroves sometimes have these too, but theirs are much smaller and more sparse. The black mangrove is the most cold tolerant of the mangrove-forest trees, extending farther northward along the coast than the others. Its flowers produce a honey that is well thought of by some people.

The other salt-forest tree is the buttonwood, which gets its name from its groups of little spherical fruit and flowers. Buttonwood usually grows farther inland than the others, often mixing with terrestrial hardwoods. Mahogany and buttonwood often grow together, for example, especially in the once-famous mahogany country of the northern shore of Florida Bay. In some of the woods near there you can see live oak, mahogany and buttonwood growing in mixed stands. Buttonwood bark is favored by epiphytes as a surface to grow on. The trees apparently live a long time, because you often see big ones, repeatedly blown over by hurricanes, lying recumbent and looping off across the ground from one set of roots to another, or even doubling back across their own trunks before slanting up toward the sun again. The wood of this species used to be the favorite fuel of people on the Keys. Expeditions to burn buttonwood charcoal and cut mahogany brought some of the earliest settlers to the remote northwest coast of Florida Bay.

In Florida the shore zone is usually dominated by red mangrove, chiefly young plants growing on ground that is periodically under water. Turtle grass and manatee grass are often associated with red mangrove, and there may also be patches of the marsh grass *Spartina*, and a sprinkling of seedlings of black and white mangroves. In meeting the ecologic challenge of life in the difficult, storm-wracked unstable zone where tidal salt water laps the edge of the land, the red mangrove has evolved into one of the most bizarre of all vegetables. Because the ecologic demands of the various kinds of seaside environments are very different, this species varies markedly in form. Some red mangroves are tall and straight-trunked; some are low-crowned domes standing on thin stilts; some spread out horizontally, and repeatedly reroot as they go, like vegetable centipedes.

Two of the fundamental adaptations that fit the red mangrove for its perilous life at the edge of the sea are salt tolerance and an ability to cling to unstable ground. A third adaptive achievement is a reproductive device that allows the species to disperse widely and colonize new habitat. Red mangroves have little yellow, waxy flowers; the seeds that these produce germinate before they leave the tree, generating a cigar-shaped seedling 6 to 12 inches long. When these seedlings fall, some may lodge and take root beneath the parent tree and become part of a growing forest there, but most of them drift away with tides and currents and may travel for hundreds or even thousands of miles before they strand or die. When a seedling strands in shallow water it quickly grows roots, and these pull the little stem erect. A few leaves appear, and in a short while a new little mangrove stands in the shallows. As it grows, thin aerial roots arch out of the trunk or drop from the spreading limbs. As each of these touches the bottom it frays out into thin claws that seize the mud to anchor the young tree against the waves, tide-wash, and winds of seaside storms. The roots multiply, the stem rises until it is lost in the growing crown, and eventually the tree may stand only on its multiple prop roots. The roots have several special functions: they bring up sap; they raise the short trunk above the water; they serve as breathing organs to furnish oxygen that is not available down in the mud; they make a system of props and braces that holds the tree erect in unstable ground; and finally they reduce wave-wash and current flow, and so promote the accumulation of silt and detritus, and consolidate the hold of the tree on the unstable edge of the land. A single, well-grown, tidal-zone red mangrove standing high on its thin legs is a strange-looking plant. A forest of such trees is one of the most offbeat kinds of vegetation to be found anywhere.

Although there is some tendency for the different kinds of mangrove-forest trees to segregate according to the salinity of the water, this is not nearly so marked in Florida as it apparently is in other parts of the world. The outermost pioneering rank is nearly always composed of red mangrove, and buttonwood is the most prevalent on drier land; but otherwise, the species mix unpredictably. The red mangrove is even able to survive in fresh water, though it seems unhappy there. One of the more striking landscapes to be seen along the road through Everglades National Park to Flamingo is a stand of dwarf cypresses that meets a stunted outlier of the red-mangrove fringe growing with the fresh-water sawgrass, under what appear to be almost intolerable ecological conditions. The tiny, many-legged, widely spaced mangroves probably grew from seedlings blown there by hurricanes, and they seem barely able to survive in the scanty soil that covers the limestone bedrock. Some of them are only a couple of feet high, but they are obviously not young trees; the botanist Frank Craighead thinks some of them may have been there for a hundred years or more.

Contrast those spidery growths with the trees of the original mangrove forests of Florida, before the tan-bark crews got to them to strip off the bark that was widely used in tanning and dyeing, and before the forests were lashed by the disastrous hurricane of 1935, then by Donna in 1960, and finally by Betsy in 1965. What must have been one of the finest of the Florida mangrove forests evidently grew about the mouth of Little River, between Fort Lauderdale and Miami. There, according to Charles Torrey Simpson, "Some of (the trees) were braced by air roots fully eighteen inches in diameter that sprung from a height of twenty-five feet above the ground, and in other cases slender roots dropped from the branches fully thirty-five feet above the soil. . . . These trees easily ranked among the most wonderful vegetable growths of the State of Florida."

Though most of the mangrove forest that remains is typically composed of lowish, round-topped trees sprawling over tidal shallows along the shores of brackish bays and estuaries, the trees rise high and straight-stemmed when they find themselves on slightly higher or deeper or more stimulating soil. I don't know what the top heights of such mangroves are around the world, but people who speak of the forests of some parts of Southeast Asia often mention 100 feet, and the same figure has been repeatedly mentioned as the maximum height of the red mangroves of Florida.

An old forest of red mangroves growing on slightly elevated solid ground can be walked through as you would walk through any dense woods. To walk through mangroves in tide-covered ground, or to wade or swim through them, is next to impossible. By far the best way to go, if this should ever be really necessary, is to scramble and brachiate—that is, swing by the arms from one hold to another—through the interlocking branches, well above the shell-encrusted basketwork made by the arching and interlacing prop roots.

Throughout the tropics, mangroves are of utmost importance in the ecology of marine estuaries, the bays or river mouths in which the sea meets the fresh water running off the land. So long as estuaries remain untouched, the fresh water they receive and the nutrients that this brings in are mixed and mellowed in an orderly way with the tidal sea water, producing what have belatedly come to be recognized as some of the richest environments in the world. Besides supporting a diverse fauna of their own, estuaries provide a breeding place and nursery ground for a great many kinds of marine animals, and they are a way station in the migratory travel of many others. The recent growth of public interest in human ecology has called attention to the widespread destruction of natural tidal-shore environments. Until lately tidal swamps and marshes have been thought of as dismal half-worlds that ought to be dredged, filled and bulk-headed to make clean-shored waterfront real estate. Florida has suffered grievously from this compulsive destruction of shoreline environments, and because the state has so much shoreline and is growing so fast, the damage has been appalling.

The air-absorbing roots of a black mangrove rise from the water in dense array. Called pneumatophores, these solid structures—about a half inch thick—grow up from the plant's submerged lateral roots and supply the tree with enough oxygen to prevent it from drowning in sea water or smothering in mud. The leafy shoot at center is an encroaching red mangrove.

Mangroves make a vital contribution to the estuarine environments of southern Florida, protecting them against the hydraulic power of hurricanes. Of all living organisms the mangroves are best able to stay in place during the earth-moving exercises of a 200-mile wind and the attendant seas. Without mangroves the big storms would continually reshape the coast in the hurricane belt. There is no telling what southern Florida would look like if mangroves had never flourished there.

The staying power of tide-zone mangroves has saved people as well as coastline. In the days when hurricane warning systems were poor —when, for instance, Florida depended heavily on bulletins cabled by a priest at a Catholic college in Havana—the deadly big hurricanes often came by surprise; and a main recourse of people who got caught was to take refuge in their boats, which they anchored firmly in narrow creeks in mangrove swamps. I have heard old conchs, as the Keys people used to call themselves, say repeatedly that the mangroves were the place to be when hurricanes came.

Besides their role in steadying the interplay between fresh waters and the sea, mangroves contribute directly to the nutrient cycles of coastal regions. The leaves of red mangroves are an important food for various insects, including several kinds of butterflies. The larva of the common mangrove skipper, for example, favors these leaves, and the larva of a species of moth lives in the reproductive seedlings. Mangrove leaves have been found in the stomachs of sea turtles, both the hawksbill and the green turtle. The stomach of a Honduras hawksbill that I once examined was tightly filled with pieces of mangrove seedlings that had been bitten off in sections about an inch long. Back in the days when green turtles were held for long periods awaiting the schooners to take them to market, they were kept in crawls—water-filled palisaded pens —frequently located on mangrove shores. The turtles were sometimes fed on red-mangrove leaves. Nobody seems to know how important mangrove detritus may be as a natural travel ration of the Atlantic green turtles, whose periodic migrations take them through regions in

which the submarine vegetation they prefer is lacking, and where fallen mangrove leaves are probably the only plant food they see. But the detritus would seem to be more than an emergency ration. You often see green turtles in mangrove creeks. During a short cruise only a few weeks ago I counted six half-grown ones in the mangrove estuaries around the mouths of the Shark and Little Shark Rivers.

Although land animals are not at all conspicuous in the Florida red-mangrove forests and swamps, a great many kinds of birds come and go in them. Gray squirrels occasionally live there at fruiting season, and fox squirrels are sometimes seen in the Collier County mangroves, which to a northern Floridian seems incomprehensibly out of character, because fox squirrels belong in long-leaved pines. Raccoons are abundant in even the most chaotically jumbled mangrove forest; wildcats and panthers travel in it to get back and forth between hunting places; and even antlered deer are able to negotiate it with ease.

I don't know how a deer can possibly get through a mangrove swamp. A tidal mangrove forest is a place of such monumental disorder that you would swear it had been designed by a demented maker of children's Junglegyms, or by a computer programed to keep out everything but snakes. Yet deer do somehow get through; and the raccoons pass without apparent hindrance. But pity the poor coonhound when the coon takes to the mangroves. Back in the days before television superseded coon hunting as the evening pastime of rural Floridians, there was much suffering by hounds whose masters hunted too near the mangrove swamp. Coons systematically head straight for the mangroves when dogs are after them, knowing better than the dogs appear to know that dogs can't travel there. I knew a coon hunter on the east coast who spent a whole night fighting his way through mangroves to where his dog had hung itself up in a tree.

You sometimes see rat snakes prowling in mangrove trees, and once in a blue moon you come upon a tree frog. In Chapter 4, I mentioned the intolerance of frogs for salt water. This applies mainly to direct contact, however. A frog can't be born in a mangrove swamp—it has to move there from a fresh-water pond or ditch somewhere; but sometimes you find mature frogs that have wandered far out into saline vegetation to reap some special harvest. On a recent visit to Everglades National Park I came upon a gangling, slim-legged, green tree frog in a tidal stand of young red mangroves on the shore of Florida Bay. The frog was a handsome one, with an ivory stripe down each side and with gold flecks scattered on its green silk back. It was industriously catching mosquitoes. One usually sees such frogs hunting on window screens at night. You rarely find one hunting in the daytime, and even more rarely in a mangrove tree. But this one was in a perfect frenzy of mosquito-catching. It was a drizzly day and there were so many mosquitoes it was hard to breathe, yet I noticed that not one of them sat on the naked frog. It seemed totally immune to the hordes, as if shielded from them by some personal secretion of bug repellent. On straw-thin hopping legs it dangled and swung from twig to twig, snapping at the inexhaustible store of prey on the glossy leaves, as seemingly at home in mangroves as if they had been willows or buttonbush trees. On the bigger Keys and all through the mainland mangrove fringe, mosquitoes are not only a seasonal nuisance but also a fundamental part of the energy cycle of the region. It would be hard to overestimate the importance of mangrove mosquitoes as a source of food for the Everglades' animals, ranging from the little fishes that catch the larvae to the bats, frogs and lizards that prey on the mature mosquitoes.

All sorts of marine creatures, both large and small, favor mangrove-bordered water. Manatees are partial to such places. Alligators and crocodiles used to come together there and still perhaps do, once in a while, in the small patch of country in which crocodiles remain. A number of kinds of fishes prefer mangrove-bordered streams and bays. Most of the numerous fishermen that cruise through the mangrove country are going after tarpon, snook, mangrove snappers or redfish; the rest of them are simply lost.

There are many kinds of crabs in the mangroves, ranging in size from the fiddlers that teem on mud flats at the edges of the forest to big blue land crabs in the higher tracts of swamp. Where the mangroves border bays or estuaries the mangrove terrapin—a southern relative of the northern diamondback terrapin cherished by gourmets—may be found. Low tide in the mangroves reveals the coon oysters, which cling to the aerial mangrove roots and hang in clusters in the open air when the tide goes out. Coon oysters are very good to eat but hard to get out of their sharp-edged shells.

Another highly edible animal that sometimes gets into the edges of the mangroves is the spiny lobster. Once when my family and I were snorkeling in the mangrove islets of Florida Bay we found the whole submarine fringe of prop-root basketry around two islets to be crowded with young spiny lobsters, ensconced in the wall of roots, and with their antennae all thrust seaward. Most of them were under legal size.

Some were not, however, and these we tried to catch, but we were never able to extricate a single one from the wall of wickerwork.

Along with the hospitality they offer marine creatures, mangroves make another contribution to the fauna in this part of the world: they provide nesting places for the herons and other water birds without which Florida would be a very dismal landscape.

A water-bird rookery is one of the marvels of the natural world. The gregariousness of wading birds, the tendency for the different kinds to fraternize on a foraging ground, is not restricted to fishing time. It is even more dramatically displayed in their nesting assemblages—in the mixed rookeries they build.

There can be no doubt that most wading birds share a common notion of what kind of place makes a good nesting site. As disorderly and unplanned as a mixed rookery may appear, the species in it are obviously there because the place has met certain specifications: good lodgment for nests, a supply of twigs for nest-building, an accessible food supply that will meet the heavy demands of the fledgling period, and relative freedom from terrestrial predators. It is probably the latter need that explains why rookeries are so frequently located over water in swamps—significantly often over swamps with raccoon-scaring alligators in them—and why in the coastal fringes in southernmost Florida rookeries are set up so regularly on mangrove islands. Without insulation from raccoons, snakes, rats and wildcats, a bird rookery in Florida would probably not be feasible.

Mangrove rookeries seem more durable than those in fresh-water trees. In southern Florida, where hurricanes charge sporadically across tree swamps and mangrove islands, a rookery site can be destroyed by the wind in a few hours. The most famous and historic of all Florida rookeries, that in Cuthbert Lake in Everglades National Park, lasted for decades, only to be almost wiped out of existence by Hurricane Donna.

Most of the common water birds nest impartially in a number of different kinds of trees; but in Florida brown pelicans, which are rare and endangered everywhere else, nest almost exclusively in mangroves. Their east coast rookeries occur mainly in black mangroves, while on the Gulf Coast they are all in red-mangrove swamp except for the northernmost colony, located on Sea Horse Island off Cedar Key.

To insert oneself silently into the dynamic, reeking, cacophonic midst of a croaking, keening, contesting, vomiting, defecating melange of superbly graceful adult birds and monstrously unfinished young in a mixed water-bird rookery is surely one of the most memorable of all wilderness experiences. But most rookeries are too frail to withstand visitors. Wood storks or great blue herons nesting in tall cypresses will put up with a tactful intrusion beneath them, but where the nests are clustered in the low trees and bushes that most smaller herons usually favor, even an unobtrusive visitor can cause dire confusion and distress, with parent birds squawking and flapping, fledglings tumbling, and watchful, avid fish crows diving down to raid the untended nests.

The only rookeries I know in southern Florida that you can get close to without doing damage are on mangrove islands. Cuthbert Island in Cuthbert Lake—what Hurricane Donna left of it—is one. Various little mangrove islands in eastern Florida Bay support nesting colonies, some mixed, some mostly pelicans, some roseate spoonbills, and these colonies seem to take no offense at the discreet approach of a boat.

Not long ago my wife and I went through Everglades National Park again, and our last stop was Everglades City at the western entrance to the park. We took the short evening cruise on Chokoloskee Bay to see half a dozen kinds of birds nesting on an archipelago of little mangrove islands. Some of the islands were separate trees and some were little clumps. All of them were round green domes, varying only in size and stringing out in a quarter-mile arc that was silhouetted against the setting sun and a bank of purplish storm clouds. And each island was the roost or nesting place of a host of water birds.

The boat drew up quietly to within a hundred feet or so of the middle of the island arc. The captain shut off the engine, and for nearly an hour we drifted there, listening to the croaking, gurgling comment of the birds and tallying their kinds. There were cormorants and five kinds of herons in the rookeries. Two roseate spoonbills came in to roost and settled separately in little mangrove trees. The white ibises had finished nesting the month before, and now were coming in to pass the night, each separate flock appearing as a dim, wavering line in the northeast, then slowly condensing into a chain of white birds. As the flocks arrived they circled to locate empty islands, then banked steeply and came to rest in the branches. As the little, dark green islands received their birds they flowered, as if with magnolia blooms, against the dark storm clouds out in the Gulf.

The Pioneering Red Mangrove

PHOTOGRAPHS BY DAN MC COY

The red mangroves rimming the Everglades along Florida Bay and the Gulf of Mexico are among the most aggressive trees in nature. When Christopher Columbus saw their like off the coast of Hispaniola in 1494 they were "so thick a rabbit could scarcely pass through," the ship's doctor reported in a letter back home to his native Spain.

Their aggressiveness is not only a matter of density; like Columbus himself they are pioneer voyagers in waters unvisited by less intrepid stock. For the red mangrove propagates itself by sending its seedlings out to sea—after they have germinated on the parent tree *(right)*. And as the human explorer is followed first by conquerors and later by settlers, so it is with the red mangrove. The seedlings often plant themselves in shoals that have no other vegetation. Once they have fortified the site—which they do by virtually creating land where there was none—other colonizing trees move in behind them—black and white mangroves, then oak and mahogany—to provide a new habitat for creatures ranging from root-clinging oysters to treetop-nesting birds.

As it coalesces, a mangrove forest acquires a lush, glossy canopy overhead; below, twisted and intermeshing trunks and prop roots form an intricate maze. The water lapping at the roots at high tide is often stained red from tannin—the natural dye in the trees' bark that gives the red mangrove its name.

Being a voyager, the mangrove appears everywhere in the tropics: in Africa, on both coasts of Central and South America, and on Melanesian and Polynesian shores. The seedlings can survive afloat for a year until they find a hospitable sandbar to light on. Some travel thousands of miles from their parent trees before coming to rest; the New World mangroves are thought to have originated in Africa.

A view from most places on the Gulf Coast of Florida will show a slender mangrove sapling here, another two or three there, and in the distance full-blown islands of varying dimensions. A visit to the same site another year may show quite different formations of trees and islands, for they are constantly being altered by autumn storms. But even when a cataclysmic hurricane wipes out a whole forest, the irrepressible red mangrove is the first plant to return to the scene and build up the land all over again.

A red mangrove tree dips toward the Gulf of Mexico, heavy with string-bean-shaped seedlings that it will soon deliver to the water below it. There some will take root; others will ride out with the tide to root elsewhere. From the falling of such seedlings come tangled masses of trees that spread to form islands like the one on the horizon.

Clusters of sweet-scented flowers—here magnified five times—begin the tree's life cycle.

The Birth of a Seedling

The life cycle of the red mangrove tree begins with a spring flowering that turns it into a blaze of yellow *(left)*. The flowers, about an inch in diameter, with cottony white nectar-filled centers that attract bees, bloom for a few weeks and then drop off, making way in the summer for a sweet-tasting berry that squirrels and birds feed on. The fruit sprout elongated green appendages *(right)* that are the red mangrove's principal distinction. These seedling shoots are a rarity among plant growths; by the time a seedling is ready to drop from the tree it is itself already a rudimentary tree.

This form of viviparous propagation—the production of living young direct from a parent—is a dramatic instance of environmental adaptation. The seeds of most trees germinate in soil; but since the saline coastal soil that supports the adult red mangrove is too hostile for proper germination, the tree holds its seeds until they are ready to grow.

The mangrove lets the seedlings go in late August or early September—storm season, when higher-than-usual tides provide the seedlings with easy transportation away from home. If they do not root below the parent in a day or two, they drift away to implant themselves elsewhere. In a few years a sapling will be yielding its own fruit, and in about five years' time an open shoal can acquire a whole new arbor of mangroves *(pages 146-147)*.

Strawberry-sized fruit follow the flowers and sprout foot-long seedlings that drop in late summer, to take root in offshore shallows.

In this developing mangrove forest, the center tree is from three to five years old; the nearby seedlings range in age from a few days to

about a year. The upright seedlings have taken root; those that are horizontal may drift off to right themselves and take root elsewhere.

Mangrove prop roots, bending in graceful arcs, serve to brace the tree, give it air and entrap debris that decomposes to provide nourishr

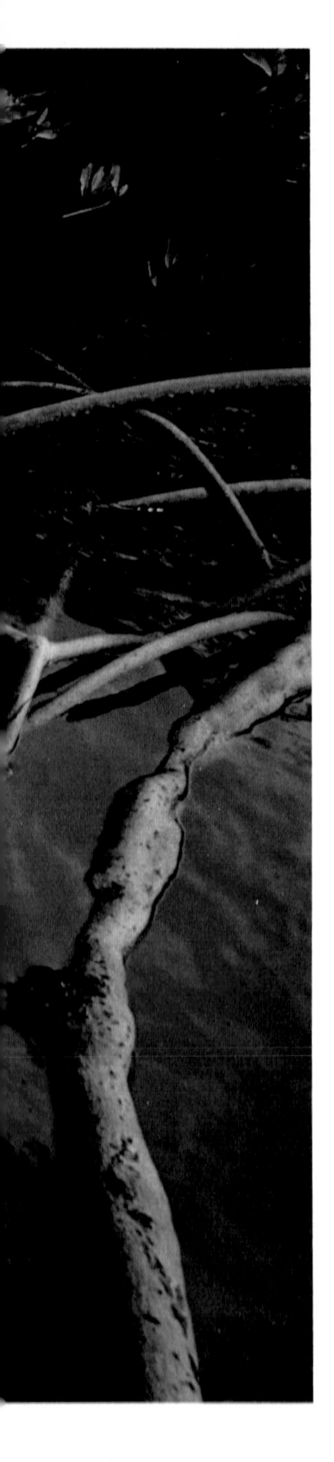

How the Mangrove Forest Grows

In its second year of life the mangrove starts to develop its major underpinnings—prop roots *(left)* that serve the dual purpose of reinforcing the tree against storms and helping transmit nourishment, moisture and oxygen from the air and soil to the trunk and branches. The prop roots are also the means by which the tree builds land; they enmesh all sorts of natural rubbish, from the leaves shed by the tree itself, to the sand, grasses and shells that are daily swept in by the sea *(right)*. As this detritus thickens and decays, it metamorphoses into soil, which further supports the growing tree and gives the advancing prop roots room and depth to perform their functions.

In a fully developed mangrove forest *(pages 150-151)* the roots form a jumble of arches so intertwined that the human eye can hardly discern where one begins and another ends. Such a forest, though hospitable to many creatures *(pages 152-153)*, is almost impenetrable to those that do not crawl or scramble, for many prop roots are too high to climb over and others are too low to stoop under.

Though mangrove forests lose ground to the ravages of storms from time to time, they have made a net gain in the past 30 to 40 years; during that period the land area around Florida and Biscayne Bays has been increased by approximately 1,500 acres, and once-open shoals have been transformed into heavily forested patches of land.

The mangrove's shed leaves are among its first soil builders.

Grasses, shells and other debris help form supportive muck.

A mature mangrove forest, with 20- to 30-foot-high trunks and a network of prop roots, fringes a Florida shore. Under ideal conditions

—meaning an absence of catastrophic hurricanes—a colony of this size and density can grow from pioneer seedlings in half a century.

Its belly bulging, a corn snake digests a meal while napping in a mangrove.

Coon oysters, a principal food of scavenging raccoons, cling to the prop roots.

A Bower for Wildlife

Every part of the mangrove forest —from the prop roots, which alternately submerge and emerge with the tides, to the topmost branches, which may tower as high as 80 feet —provides shelter or food for a multitude of creatures. This population ranges from the tiny sand fly flitting among the leaves to the 250-pound tarpon lurking offshore.

Of the bird species that find homes among the mangroves, white pelicans appear in such abundance that during the winter the trees look snow-covered from the air. Their cousin, the brown pelican *(far right)*, rare and endangered everywhere except Florida, shares the rookeries in neighborly amity with egrets, herons, wood storks, ospreys, bald eagles and cormorants.

A host of other creatures come down from higher land to forage. Raccoons favor the oysters that live off the prop roots *(left)*. Sir Walter Raleigh, exploring the New World for Queen Elizabeth I in the late 16th Century, also found them agreeable, reporting that they were "very salt and well tasted." Snakes slither up the tree trunks after birds' eggs and nestlings; cormorants, which dine chiefly on the fish in the nearby waters, even the score by catching a napping snake now and then. Fiddler crabs scuttling about in low tide perform a large service: aerating the soil as they probe for food, they increase the supply of oxygen to the trees that attract all the creatures.

A wary cormorant dries its wings on a perch in a mangrove treetop.

A brown pelican rests on a low branch between dives for fish.

A Violent End, a New Beginning

The equinoctial storms of autumn, with their especially high tides, are for the most part a blessing to the mangrove tree. They bring it nutriments that it needs for its well-being; they carry its young away to establish the beginnings of new mangrove colonies in other places.

But every quarter century or so an epic hurricane comes like a curse instead. It hurls winds close to 200 miles an hour across the land, tearing the bark from the trees and exposing their tender cores to the desiccating salt spray of the sea. Such a storm churns the waves to heights of 20 feet; on Florida's flat terrain this can cause the waters to spill inland more than 10 miles. The backwash sucks out the supporting soil from among the prop roots of the mangroves and the winds then fell even the giants of the forest. Of the major hurricanes that have ravaged Florida during this century, one, in 1960, destroyed from 50 to 75 per cent of the mature mangroves along the Shark River *(right)*, until then the site of some of the tallest mangroves in the world.

But the mangrove tree has a remarkable ability to regenerate. It sprouts new life in the very center of a hurricane's devastation. When the storm subsides, gentler tides return and deposit seedlings gathered from healthy trees elsewhere—and so begins again the process that keeps the shape of the Florida shoreline ever new and ever changing.

The dead hulk of a mangrove felled by Hurricane Donna in 1960 sprawls in the Shark River, among seedlings that presage a new forest.

7/ A Wilderness Besieged

No state is under greater pressure from all the forces that place demands upon land, water and life. . . . The United States begins or ends in Florida. RAYMOND F. DASMANN/ *NO FURTHER RETREAT*

You used to hear an unkind definition of a government African expert as a man who had made a flight over the continent of Africa. The gibe was no doubt justified, but actually if you want to visualize the anatomy of a region in a short while the best possible means of travel is by a plane flown sympathetically, searchingly and at an elevation from which one kind of vegetation can be told from another.

I made such a flight over the whole southern tip of Florida not long ago, and though I had covered most of the ground before by car, boat, airboat or small plane, nothing ever put the country together for me like that journey in a jet. The flight was one of the field trips offered at the Second National Biological Congress at Miami Beach. When I saw "Overflight of South Florida" on the list of excursions, my first thought was that I had flown over the place a hundred times already. But then I saw the flight plan they had, with a jet snaking through the whole southern end of the peninsula 1,000 feet up, missing no major feature of the country, and with Dr. Frank Craighead, consultant for Everglades National Park and the ranking authority on the region, explaining what was being flown over. I quickly signed up for my wife and myself.

The plane climbed out over western Miami into the half-world of suburbs, farms and limerock mines where the city is spreading into the Everglades. As we went into a slow southward turn I could barely make out the thin line in the west where Florida Highway 27 emerges from farms and mango groves, crosses the Tamiami Trail, and angles off northwest toward Lake Okeechobee. The road was visible so far away only because of the outlandish strips of cajeput, Brazilian peppers and casuarinas that grow along its shoulders.

There are more ways to pollute a landscape than by loading it with sewage, smog and beer cans. You can load it with exotic plants and animals. Whether brought in intentionally or accidentally, these are likely to change the landscape. The changes they make are rarely good, and often are atrocious. The droll armadillo, for instance—introduced into Florida only a few decades ago—is modifying the woodlands by its rooting for the small animals that inhabit the leaf mold. The walking catfish, which caused national excitement when it established itself in Florida a few years ago, is apparently doing no harm so far; but it may, once it gets the feel of the land. And so may any of the foreign birds, fishes, reptiles, insects and plants that now number in hundreds.

The most conspicuous changes produced by exotic organisms in southern Florida are the brand-new types of landscape dominated by the three foreign trees I mentioned: cajeput, Brazilian pepper and casuarina. The most widespread are the casuarinas, brought long ago from Australia, and now driving out or inhibiting many other trees. Cajeput is a handsome Asian relative of the eucalyptus, with cream-smooth bark that curls off in thin layers like birch bark. Being fireproof, and catholic in its habitat selection, it has invaded thousands of acres of cutover flatwoods. Cajeput tends to form biologically sterile communities, practically devoid of animal life—perhaps because the secretion the tree produces is offensive to animals. Local naturalists tell me that when cajeput comes in around a gator hole the gator will move away. The Brazilian pepper, an attractive ornamental tree or shrub with bright red berries that appeal to birds, is now growing in spreading copses all through the frost-free parts of the state.

I started to regale my wife with a small jeremiad on the subject of the new plants and animals that have been brought to this part of the state; but that seemed a bad way to start the flight, and I made up my mind to try to do better—to concentrate on happier sights below our windows.

This was not easy. From the plane, U.S. Highway 1 was a sharp line running down the Rock Rim. Greater Miami was crowding south, shoving aside the last of the pinewoods and hammocks and flooding into country where bare limerock has been smashed into truck farms.

The limestone floor of southern Florida is profoundly involved in the ecological organization of the landscape, both natural and urban. The

Rock Rim interrupts the eastward flow of surface water to the Atlantic, and impounds the low head that sets the River of Grass off in its creeping flow southwestward. The peculiar properties of this floor of fissured rock, covered and calked by muck, bear as much on the welfare of man in Miami as they do on otters and Everglades kites. The rock is the aquifer—the underground water storage and delivery system—for the Gold Coast's wells. The recharging of this system depends on the capacity of the muck to hold water. The muck in turn is the debris of an integrated biological community that is dependent upon protracted flooding. As drainage, farming and fire destroy this covering of muck, the annual dry season lengthens, the sporadic droughts become more severe, the fauna thins out in the River of Grass; and to the east of it, in the cities of the Rock Rim, the salt water rises in the fresh-water wells. There could be no more graphic proof of the interdependence of man and nature than the predicament of the metropolitan lower east coast of Florida. There, for decades, agricultural and urban development has been progressively destroying the hydrologic system that makes life possible both in the Everglades and in the cities. The realization of that fact is the greatest single hope for saving any wilderness in southern Florida, and perhaps for saving the cities themselves.

Just below Miami the plane crossed Black Point, so called for its fringe of tall dark mangroves, and turned eastward over Biscayne Bay to give us a look at Elliott Key. I knew my plan to stay cheerful would receive a setback there. Elliott Key is near the northern end of the archipelago of the upper Florida Keys and is the nucleus of the Biscayne National Monument, which is being developed to save the superb tropical reef-and-key country between Miami and Key Largo. Elliott Key is covered by tropical hammock, a dense low forest of diverse West Indian trees. The vegetation is not virgin; the island once was largely under cultivation. But even though the forest is second growth, it has had half a century or so in which to regenerate, and has come back in a continuous even stand, and in what probably is close to its original diversity of species.

"Notice the clean strip bulldozed down the center of the island," said Frank Craighead over the intercom, "right through the heart of the hammock. It came close to destroying the last known stand of Sargent's palm in Florida." Sargent's palm, known also as cherry palm, hog cabbage or buccaneer palm, is a West Indian species that grows to a height of about 25 feet, and looks a little like a small royal palm. Everybody crowded to the windows and looked down at the wide band cut straight through the woods. A wave of muttering spread through the cabin. All up and down the aisle people who had never heard of Sargent's palm asked each other what kind of vandal would wreck a piece of rare wild country that way. Then the answer started passing from seat to seat: "To kill the preservation scheme—to ruin the key as a natural area and make the conservationists give up the project."

That marred group morale badly—until somebody passed the word that the bulldozer-happy chaps were being sued for the expense of replanting the strip, and for all the costs involved in growing a mature stand again. That cheered us up a little.

Back over the mainland, we headed for the tip of the peninsula and the mangrove shores of eastern Florida Bay. On the way we flew by the infamous C-111, the "Aerojet Canal," built in 1964-1965 to provide access to the sea for the barges of a solid-fuel plant—that has since shut down. Frank Craighead summarized the complicated story of that bloodletting exercise and its threat to the vital eastern section of Everglades National Park and the ecology of Taylor Slough.

The slough is not clearly discernible from the air as a topographic feature. It is a linear complex of ponds, marshes and sawgrass glades in a broad, shallow depression of the limestone floor of the eastern Everglades. In the wet season it carries water southward into Florida Bay, flowing into Everglades National Park, across the Hole-in-the-Doughnut—the private agricultural holdings that scar the eastern section of the park—and into the midsection of the Florida Bay shore. In former times, in the wet season, small boats could be paddled all the way from Homestead through Taylor Slough to Florida Bay. In recent decades, however, with all the farming and development at its headwaters, the slough has dried out. The uproar over the recent digging in this region was stirred up by the probability that the canal complex involved would further lower the slough's headwaters. This would increase the duration of drought and the threat of fire, make the eastern section of the park more vulnerable to invasion in the dry season by salt water from Barnes Sound, and disrupt the ecologic stability of this section of the park. But court orders, generated by efforts of the National Audubon Society, have lessened the danger. Gated culverts have been installed in C-111 to permit the flow of water into Barnes Sound but prevent salt water from backing up into Taylor Slough. And work on auxiliary canals of the system has been halted.

Soon after passing C-111, we were over the zone of northeastern Florida Bay where salt and fresh water meet. From the air the irregular mosaic of lagoons and saline vegetation looked as if buckets of paint—tan, beige, chartreuse and olive—had been poured together on a perfectly smooth surface. We crossed Taylor Slough where it spreads into Little Madeira Bay, famous for its crocodiles, mahogany and mosquitoes, and for a while continued on west along the wavering saline zone with its broad band of intermingled glades and salt-tolerant vegetation. Then we turned southwestward, passing Coot Bay and coming out abruptly over Cape Sable, an obviously separate and very different kind of country, cut off from the Everglades by the vast lagoon known as Whitewater Bay. We circled over Flamingo, once a remote little fishing village, now site of the southern headquarters of Everglades National Park. Looking down on that tactful intrusion of organized humanity, I thought of my own earliest visit there with my brother in the early '30s. The first living thing we had seen at the end of our five-hour trip in a Model A Ford down the ruinous road of those days was a bobcat in a garbage can behind a tall, teetery, weather-beaten house. The rear end of a bobcat it actually was, sticking out of the can he was foraging in. I even recalled what the lady of that house said when she gave us a drink of water and we spoke of the bobcat. "They're a nuisance around here," she said. "They eat our house cats so bad we can't keep none."

I also remembered seeing a man skinning a crocodile at the old Coot Bay alligator camp, and I recalled losing a lot of plugs to baby tarpon that jumped around in a narrow creek like crazy and threw our lures up into the mangroves. Most vividly I thought of the mosquitoes, of driving back to Homestead after dark with a broken spring at three miles an hour the whole way. For most of the trip back my brother sat out on the hood of the car, in the wan hope that the whip of the overhanging trees would brush some of the mosquitoes away. But as it turned out the trees were mostly poisonwood, and the next day my brother's eyes were swollen shut and his face was like a pumpkin.

Flamingo is located at the bottom of Cape Sable, which actually consists of three subcapes: East Cape, Middle Cape and Northwest Cape, connected by surf-built beaches. The best beach is at East Cape, the southernmost projection of the mainland United States. There is an important nesting colony of the loggerhead turtle on Cape Sable. Along with all other kinds of sea turtles, the loggerhead is declining the world over. Its reproduction is increasingly hindered by the loss of wild seashore and by the egg hunting that still goes on in parts of the nesting range. Although Cape Sable suffers frequent hurricane damage it is good loggerhead nesting shore, and because the beaches there are bet-

Thick with tropical hardwoods, Elliott Key predominates in this air view looking north at the keys of upper Biscayne Bay. The zigzag line down the middle of the key is the remnant of a roadbed that was bulldozed in 1968 as a prelude to real-estate development. The project was thwarted when the island was included as the nucleus of the Key Biscayne National Monument, and the abandoned roadbed is now being reclaimed by tropical vegetation.

ter shielded from human intrusion than most of the mainland coast, they deserve careful attention as a sea-turtle sanctuary.

The only serious natural menace to the Cape Sable turtles is the raccoon. Coons love turtle eggs, and at Cape Sable the coons are so abundant that it is hard to see how the loggerhead nests there. The big coon population may not be natural, but another sign of man's disruption of ecological balance, in this case by his reduction of the panthers and alligators, the two most important coon predators.

A conspicuous feature of Cape Sable is a broad crescent of marl prairie that spreads inland behind the beaches. These prairies are not residual sea bottom, but wave-built land. Although a slow rise of sea level has occurred at the cape during the last 5,000 years or so, at the same time the land has been repeatedly built up by wave-deposition during the sporadic hurricanes. That process has created a series of ridges, which can clearly be seen from the air, running parallel with the shore. This slightly elevated land makes a barrier to the flow of the Everglades' water and diverts it westward into the Gulf of Mexico.

To keep the geography in view, our pilot held a course straight up the cape, flying along Lake Ingraham, on the prairie behind the clean ocean beach, and passing over old canals that mark the dashed hopes of early real-estate men. We turned out over the Gulf again at Middle Cape, where a ragged stand of palms is all that remains of a huge coconut plantation that was set out in the 1880s and soon abandoned.

The land beneath us was as wild as any in Florida. As far as we could see in any direction the old coco palms were the only sign that man had ever tried to tame it. From the days of the earliest explorers the conspicuous geographic position of Cape Sable, jutting into the Gulf, brought it to the attention of travelers. But the cape escaped the pandemic of development because of its remoteness—and its mosquitoes.

Thank the Lord for the mosquitoes. The world owes them a lot for their part in preserving Cape Sable. A heroic statue of a mosquito in bronze ought to be set up on a hurricane-proof pedestal, a huge plinth of Key Largo limestone perhaps, at some commanding point on the cape. There might be a little trouble deciding what kind of mosquito the statue ought to represent. As I understand it, three kinds have borne the main burden of keeping people out. One is *Psorophora*, known locally as the Everglades mosquito. Another is the salt-marsh mosquito or, as the conchs call it, mangrove mosquito—*Aedes taeniorhyncus*. The third is *Aedes sollitans*, commonly known as the New Jersey mosquito.

Apart from the deterrent influence of mosquitoes on development, they deserve attention as one of the really imposing biologic phenomena of the region. Their abundance at certain seasons—I think of June with most uneasiness—surpasses belief, or used to. Madeira Bay stands out in my memory as the mosquito capital of the world; but the whole crescent from Shark River down the cape and around the shore of Florida Bay to Key Largo probably has more mosquitoes than any place on earth. People from New Jersey, Labrador or Alaska will tell you that nobody has seen mosquitoes till he visits their area; but this must be mere regional chauvinism. The abundance of mosquitoes in June at Madeira Bay is controlled only by the amount of space there is to hold them.

Madeira Bay mosquitoes do more than punch countless holes in your skin—they interfere with your breathing. There are authenticated cases of people succumbing to mosquito attack in former days. In telling of these events it is customary for the narrator to say, "Sucked him dry, poor devil." But according to James S. Haeger of the Entomological Research Center at Vero Beach, the mosquitoes probably did more than that. In his view, while loss of juices might have been a factor, reaction to the saliva a mosquito injects probably made trouble also, and there was also probably respiratory damage resulting from both mechanical obstruction and the irritation that the wing scales of mosquitoes produce in mucous membrane. When you add to all this the panic that sometimes overcomes the victim—as it did one man who had a flat tire on Key Largo one night and tried to run to Homestead through a cloud of salt-marsh mosquitoes—no wonder fatalities have occurred. The wonder is that in the old days there weren't more victims, with railroad workers getting drunk on rotgut on Saturday nights and passing out on the roadside. One of the standing mysteries of Florida is what took place in the biochemistry of the Indians, when they came down into this country, that let them live without screens or bug spray.

A short way beyond Northwest Cape our plane banked and headed northeast, moving over the complex of sloughs and mangrove islands around Ponce de Leon Bay. To the degree that the River of Grass is, in more than a metaphoric sense, a river, this country is its mouth and delta. It is obviously a drowned shoreline, with no beach anywhere in sight, only eroding islets partly stabilized by red mangroves.

As we zigzagged up the Shark River distributaries to Tarpon Bay and on into the lower Shark River Slough, the mangroves slowly dropped out and were replaced by the Pahayokee landscape, the grass-and-tree-island country of the typical Everglades. The hundreds of little separate, insular heads and hammocks set about on the sawgrass plain were mainly elongate-teardrop in shape, with the big ends upstream and the thin ends drawn out toward the southwest. The shape is supposed to have been molded by running water, no doubt during times when currents flowed much stronger than they do now. Most of the tree-islands also have a rock outcrop at the upper end. Some people believe that all of the larger islets in this lower section of the Shark River valley were inhabited by Indians—the aborigines, that is, not the Seminoles; there are more tree-islands in the Glades than there ever were Seminoles in southern Florida.

Heading on up the valley, we flew over the Shark Valley Observation Tower and followed the tower road up to the Tamiami Trail, crossing it just east of Forty-Mile Bend—the dogleg made when the two construction crews, one building out from Miami, the other from the west coast, failed by a bit to come together. From there it was only a moment before we were over the vast, raw rectangular patches of bulldozed lifelessness where a major catastrophe had lately been so narrowly averted when the federal government halted further construction on a training-flight strip at the planned site of an international jetport. It was good to leave this unhallowed place behind and cruise out over the Big Cypress Swamp into Seminole country, and to hear Frank Craighead saying that we were flying over the traditional site of the Green Corn Dance, the major springtime ceremony of the Indians.

Turning back southwestward over the swampland, we moseyed out along the Loop Road, the old route to the settlement of Pinecrest, which runs through a marvelously rich and varied country where cypress swamp, bayheads, hammocks and pineland alternate with tracts of sawgrass glade. All through this part of the country the apparition of a Boeing 727 a thousand feet up kept stirring up flocks of white water birds. I couldn't make out whether they were common egrets, snowy egrets, cattle egrets or immature little blues; but they were there in comforting numbers. The westerly course took us back into the salt zone once again, the cypresses gave way to salt marsh and ponds bordered with mangrove, and flocks of water birds became so frequent you could almost fancy you were looking down into the Everglades of the 1930s. Then we crossed a broad belt of mangrove swamp and suddenly came out over the Gulf of Mexico and the Ten Thousand Islands.

From the air this is the most striking landscape in Florida. It is a mosaic of separate mangrove islets, spaced evenly but widely random in their shapes. The outer ones have white sand beaches; the rest are clean-edged patches of what looks like green felt, laid out like a vast jigsaw puzzle that has been assembled but has all its pieces pulled evenly apart. This bewilderingly redundant archipelago begins at Cape Romano and extends southward for some 20 miles to Pavilion Key, where it gradually merges with the slightly less wildly convoluted mangrove patchwork that continues to Cape Sable.

We circled out over the Gulf and came in again over Chokoloskee Island, once an important community of the Calusa Indians, who predated the Seminoles. The whole island is essentially a shell mound, covering some 135 acres and 20 feet high—the highest land, except for Marco Island, anywhere in the Ten Thousand Islands. After circling over Chokoloskee we headed inland, and Everglades City appeared below the plane. This is the western entrance to the national park. The ranger station there is where you record your travel plan if you aim to cruise the marked boat route of the 100-mile-long Wilderness Waterway to Flamingo. You have only to look at the twisting route from the air to see why telling somebody your travel plans is a good precaution to take.

Beyond Everglades City the spartina marsh grass and mangroves gave way to sawgrass sprinkled with little cypress trees, and then to big cypress heads and strands. Cutting back over the Tamiami Trail, we were quickly over the little town of Copeland and from there over the Fahkahatchee Strand, the main drainage slough of the Big Cypress Swamp. After some searching the pilot found Deep Lake, located in Deep Lake Strand, which runs east of the Fahakatchee. Deep Lake, a lime-sink pond 95 feet deep, is the deepest natural body of fresh water in southern Florida. It is a solution sink formed by surface water percolating down through fissured limestone that once stood much higher above sea level and ground-water level than the present lake does.

From there we turned toward the Gulf again and into a ghastly gridwork of canals where a vast real-estate project was recently laid out in uninhabitable swampland that in the long run would have been worth more as intact wilderness than all the assets of the developers involved in ruining it. When we came out over the Tamiami Trail again, we could see where the spillway of the main canal running alongside that melancholy undertaking was bleeding the water of the Big Cypress Swamp out to the Gulf of Mexico by the millions of gallons a day.

Northward we flew to Marco, largest of the Ten Thousand Islands, and now prominent as the site of the state's most recent and most imposing instant metropolis. Marco Island is about six miles long and three miles wide. It was built up by the combined action of storm seas building sand dunes and of Calusa Indians accumulating clam shells. One archeological site there, excavated by Frank Hamilton Cushing in 1895, yielded a diverse array of incredibly well-preserved wooden masks, tools and ornaments. The material came from muck-filled areas between a series of shell ridges, and was preserved by the water-soaked muck that covered it. The articles were made of a number of kinds of wood, including pine, cypress, mangrove and other species not yet identified. The trees were felled and cut into lengths with shell axes, then shaped with adzes made of sawfish teeth, scrapers made of barracuda jaws, or chisels made from the heavy columellas—the central spines—of big conch shells. The smooth finish of some of the articles suggests that they were sanded with sharkskin.

The beds of huge quahog clams that once thrived in the surrounding waters were evidently one important reason for the Calusas' being there. The clams were still extant when I first visited Marco in the '30s and were, in fact, the main reason I went there. Soon afterward they were wiped out, probably because of overexploitation.

As we moved in over the $500-million development that now spreads over Marco Island I remembered the strange little lost-looking salt-water hamlet that once was there, the gray-shingled, pile-supported houses with racked-out fishing nets before them; the clam factory; and the quiet, ponderous quahogs, bedded like cobblestones in the shallows, straining the pristine water of those Depression days as they had when the oldest Indian shellfish cultures thrived there. Suddenly it felt curious to see 60 people bending their necks to look down through the windows of a jet on a dozen square miles of streets and urban canals that seemed laid out to test the maze-running wits of the inhabitants of the fine homes spaced out along them. If there is one point of sharpest focus for the agonizing problems southern Florida faces, it is in that luxurious city that has arisen overnight by developers' fiat in the utterly remote jumping-off place that the Ten Thousand Islands once were.

After meditating on the numbing spectacle of Marco Island for a while, we left it and headed northward. We flew along nine miles of uninhabited Gulf beach, marred only by a continuous strip of volunteer casuarinas, which have the habit of springing up in new seaside groves when seeds wash ashore with hurricanes. We passed offshore of Rookery Bay where, by a famous *tour de force* of regional planning, a wilderness of mangroves, fishes, birds and water has, for the time at least, been saved at the edge of metropolitan Naples. We loafed around over the Gulf looking at the Naples skyline while Frank Craighead told us how the people there created a worldwide stir when they made of the Rookery Bay project a model of man-land harmony.

The project came about when people began tossing in their sleep over dreams of Naples and Marco running together. The long-range scheme for the region included a highway down through the Ten Thousand Islands, and south of that a causeway across Florida Bay to Key West. That was for the future; but explosive development between Naples and Marco was imminent, and it was to try to control this that the Collier County Conservancy was organized. This extraordinarily forceful group defeated a move to open Rookery Bay to automobiles. With the help of the Nature Conservancy of Washington and the National Audubon Society, it bought private lands with which to buffer the 1,500 acres of public lands in and around the Bay, and the 4,100-acre Rookery Bay Sanctuary was formed. But to protect the delicate ecologic balance within the sanctuary, control over the whole landscape was vital; and in 1967 the Conservation Foundation embarked on a program of research and public relations to bring this about. As auspicious as the Rookery Bay project appears, the area is by no means ecologically self-contained. If in the long run the area is somehow to be kept clean and wild—a place where young sea fishes grow and where quiet people can watch waterfowl tend their clamorous nests—this will clearly have to be done under a growing siege from mankind and mankind's works.

In a nest dug on a Cape Sable beach, a loggerhead turtle lays her eggs (above), generally numbering 100 or more, then uses her flippers to cover them with sand (right). The infants, which hatch within two months and take to the water immediately, are two-inch-long miniatures of their ponderous parents, which may weigh up to 500 pounds.

North of Naples we flew along more miles of good beach, undeveloped for the nonce, though there were casuarinas behind it; and then we turned inland to Corkscrew Swamp. Corkscrew is the only sizable tract of the Big Cypress Swamp that remains uncut and, as I noted in Chapter 2, the Corkscrew Sanctuary harbors the most important remaining rookery of the beleaguered wood stork. Located north of the mayhem that developers are committing in the drainage pattern of much of the southern part of Big Cypress, Corkscrew seems at the moment to be one of the most stable samples of original landscape under preservation anywhere in southern Florida.

From Corkscrew we skimmed over mixed cypress swamp and pinelands to the town of Immokalee, which stands 40 feet above sea level —higher than any place south of Lake Okeechobee except Marco Island. All the land is laid out in the rectangles of farms and ranches.

A short way to the northeast we entered the strange spread of coun-

try known as the Devil's Garden, a wild hodgepodge of flatwoods, hammocks, bayheads and water that suddenly here and there gives way to a geometric landscape of circular ponds strung out in patterns of straight lines. Many of the ponds are rings of water surrounding central islands. The geologic origin of Devil's Garden seems not to be known. Somebody suggested that a shower of meteorites made it, as they are said to have made a somewhat similar landscape in North Carolina. The region is probably some special kind of karst topography—a sunken limestone landscape, in which the solution ponds follow the course of old surface or subsurface drainage flows.

I had driven through the Devil's Garden several times, but this was my first good look at it from the air, and I marveled that such dramatic terrain should all the time have lain out there among the little roads I had traveled on so often. Our altitude was just right to show off the geometry of the place and the flocks of herons that our crossing sent wafting about the countryside; we were also able to tell most of the kinds of trees apart. This strange land is now being rapidly lost to pasture, and as with most of the original biological landscapes of Florida, nothing has been done to save a sample of it.

Traveling on over pines, palm hammocks and flocks of flying water birds, we flew north into vast sugar-cane lands, on to the sugar town of Clewiston, at the southwestern shore of Lake Okeechobee, and out across that great inland sea. Along the eastern shore of the lake once stood one of the finest stands of tropical hammock in the state. It grew on a deep peat soil that proved its undoing soon after white men got there because it grew beans and cane too well. Two thirds of the way down the shore we turned southeastward over the lakeside town of Pahokee—the name is a variant of Pahayokee, the Seminole name for the sawgrass glades—and followed the West Palm Beach canal across ranches, farms and fruit groves into Water Conservation Area No. 1. This is the northernmost of the three big impoundments designed to control the water supply of the area. Farther southward we crossed Conservation Area No. 2A, a place notable for the huge catches of largemouth bass that for some reason are made there from time to time.

We traveled eastward till Fort Lauderdale was under us—the city limits creeping westward into the Glades at a pace you could almost see. Then suddenly we were out over the Atlantic and the Gulf Stream, where charter boats towed the snow-white v's of their wakes across the indigo water. From there it was no time till we were streaking in toward the fabulous façade of Miami Beach, with the city behind it, its solid back to Biscayne Bay, and more city across on the far shore and on out into the western noontime haze as far as you could see.

With our flight nearly over, it seemed an appropriate time for more reminiscence, and I thought of the 1890s when Lieutenant Hugh Willoughby, about to begin a canoe trip across the Everglades, described Miami this way:

"What a change has been made in this place since the same time last year! . . . Of course, its splendid big hotel [The Royal Palm], with every modern convenience, will prove a great boon to the tourist, but for me the picturesqueness seemed to have gone; its wildness has been rudely marred by the hand of civilization.

"In all Florida I have never seen a more beautiful spot than where this deep, narrow river [the Miami River] suddenly opens into Biscayne Bay. . . . Of course it will all be very beautiful around the hotel. . . . I regret the change. . . . But in the nature of things the wilderness must be gradually encroached upon. . . . We must not look upon these things from the sentimental point of view. The romance and poetry must be suppressed for the sterner, material welfare of our fellow-man."

As we turned into our approach to the airport, I wondered idly what Willoughby would have thought about the balance of poetry and material welfare down there below us, and then my thoughts turned to the little burrowing owls that live in holes in the grass flats among the runways, and it occurred to me that I ought to find out how they had fared when the Boeing 747s first came to this airport.

Then we touched down, and a few minutes later the engines stopped, just where our trip had started 2 hours and 20 minutes before. When the doors of the plane opened no instant authorities on the southern Florida ecosystem emerged, but 60 people came out having seen a complicated country, with a treasure of wild creatures and wild landscapes, and with a heavy load of problems. To me the low-flown reconnaissance brought a new feeling of the oneness of the region. I went away confident that the choice that southern Florida faces is not between water for birds and water for people, as short-sighted boosters were proclaiming a little while ago. The question is, rather, whether both shall survive on a shared water ration in a magic but pitifully fragile land.

The Storm-haunted Shore

Like an ancient battlefield, Cape Sable at the southwestern tip of Florida, in the track of hurricanes, is a lonely place haunted by a history of violence. It is the cape's fate to keep reliving its turbulent past—to recover from one hurricane only to face another. And the awful, destructive force of tropical storms has turned this peninsula into a strange patchwork of desert and swamp.

The storms hit the cape hardest along its Gulf shore, a 30-mile arc of broad beaches broken by three "V"-shaped spits of land known as East Cape, Middle Cape and Northwest Cape. In many areas along this coast, stormy seas chew away great chunks of beach, indenting the shoreline by as much as 100 feet. Elsewhere waves build up huge mounds of sea shells, thick blankets of mud and seaweed and piles of shattered driftwood. Here and there lie young mangrove trees undermined by the waves or felled by winds of 100 miles an hour or more. Mature mangroves usually are strong enough to hold their ground, but many are stripped of their leaves and bark, enduring like skeletons in ghost forests behind the battered beach.

Farther inland, alternating belts of sand and swamp present curious proof of the storms' shaping force: as giant hurricane waves spill into the low-lying interior, the sand they sweep with them forms a low ridge that traps sea water in a trough on the inland side. The resulting brackish pools become overgrown by mangroves and other swamp vegetation. And growing incongruously on the dry sand ridges between the swamps are desert plants—agave, yucca, cactus—and cabbage palm. During the worst hurricanes, when waves sweep all the way over the peninsula, some of these plant communities are destroyed. But always new vegetation springs up in the same alternating patterns. And when pioneering red mangroves sprout at the edge of an eroded beach, the soil collected and created by their dense roots gradually extends the shoreline back out to sea.

The cape's recovery from a hurricane is always painfully slow. But as a matter of principle, the rangers of Everglades National Park—of which the cape is a unique part—make no attempt to hasten the process by man-made means, lest they risk altering the natural course of regeneration. So the cape is left wild and desolate to work out its own tempestuous evolution.

A DESOLATE BEACH AT MIDDLE CAPE

ROOTS OF DEAD MANGROVES EXPOSED BY EROSION AT NORTHWEST CAPE

A·CABBAGE PALM RISING ABOVE A CLUMP OF YUCCAS (RIGHT)

SPINES EDGING AN AGAVE LEAF

A MASS OF SEA SHELLS NEAR EAST CAPE

BRANCHES OF A WIND-STRIPPED MANGROVE

MOUNDS OF SHELLS PILED UP BY THE WAVES AT MIDDLE CAPE

Bibliography

*Also available in paperback.
†Available in paperback only.

*Bent, Arthur Cleveland, *Life Histories of North American Marsh Birds.* Peter Smith, 1963.

Carr, Archie, and Coleman J. Goin, *Guide to the Reptiles, Amphibians and Fresh-Water Fishes of Florida.* University of Florida Press, 1959.

*Caulfield, Patricia, *Everglades.* Sierra Club, 1970.

Conant, Roger, *A Field Guide to the Reptiles and Amphibians of the United States and Canada East of the 100th Meridian.* Houghton Mifflin Company, 1958.

†Craighead, Frank C., *Orchids and Other Air Plants of the Everglades National Park.* University of Miami Press, 1963.

Craighead, Frank C., *The Trees of South Florida, Vol. I: The Natural Environments and Their Succession.* University of Miami Press, 1971.

Dasmann, Raymond F., *No Further Retreat–The Fight to Save Florida.* The Macmillan Company, 1971.

†Davis, John H., Jr., *The Natural Features of Southern Florida.* The Florida Geological Survey, 1943.

Dimock, A. W. and Julian A., *Florida Enchantments.* The Outing Publishing Company, 1908.

*Douglas, Marjory Stoneman, *The Everglades–River of Grass.* Rinehart & Company, 1947.

Federal Writers' Project of the Work Projects Administration, *Florida–A Guide to the Southernmost State.* Oxford University Press, 1939.

Ford, Alice, *The Bird Biographies of John James Audubon.* University of Oklahoma Press, 1965.

Gantz, Charlotte Orr, *A Naturalist in Southern Florida.* University of Miami Press, 1971.

Hanna, Alfred Jackson and Kathryn Abbey, *Lake Okeechobee–Wellspring of the Everglades.* The Bobbs-Merrill Company, 1948.

*Harrar, Ellwood S. and J. George, *Guide to Southern Trees,* 2nd ed. Peter Smith, 1962.

†Hawkes, Alex D., *Guide to Plants of the Everglades National Park.* Tropic Isle Publishers Inc., 1965.

Howell, Arthur H., *Florida Bird Life.* Coward-McCann, Inc., 1932.

*Johnson, James Ralph, *The Southern Swamps of America.* David McKay Company, Inc., 1970.

Kennedy, Stetson, *Palmetto Country.* Duell, Sloan & Pearce, 1942.

Longstreet, R. J., ed., *Birds in Florida.* Trend House, 1969.

McIlhenny, E. A., *The Alligator's Life History.* The Christopher Publishing House, 1935.

Pough, Richard H., *Audubon Water Bird Guide–Water, Game and Large Land Birds; Eastern and Central North America from Southern Texas to Central Greenland.* Doubleday & Company, Inc., 1951.

†Robertson, William B., Jr., *Everglades –The Park Story.* University of Miami Press, 1959.

Safford, W. E., "Natural History of Paradise Key and the Near-By Everglades of Florida." *Smithsonian Report for 1917.* Government Printing Office, 1919.

Sanger, Marjory Bartlett, *Mangrove Island.* The World Publishing Company, 1963.

Simpson, Charles Torrey, *In Lower Florida Wilds.* G. P. Putnam's Sons, 1920.

Tebeau, Charlton W., *Florida's Last Frontier–The History of Collier County,* rev. ed. University of Miami Press, 1970.

Tebeau, Charlton W., *A History of Florida.* University of Miami Press, 1971.

*Tebeau, Charlton W., *Man in the Everglades–2000 Years of Human History in the Everglades National Park,* 2nd ed. University of Miami Press, 1968.

Acknowledgments

The author and editors of this book are particularly indebted to Donald Goodman, Assistant Professor of Biology, University of Florida, Gainesville. They also wish to thank the following persons at Everglades National Park, Homestead, Florida: Jack E. Stark, Superintendent; Ralph E. Miele; John C. Ogden; Robert L. Peterson; John Wesley Phillips; James Sanders; William W. Shenk; James E. Watters. At the University of Florida: Oliver Austin, Professor of Zoology; Thomas Emmel, Assistant Professor of Zoology; Daniel B. Ward, Associate Professor of Botany. At the National Audubon Society: Sam Dasher; Louise Donohue; Alexander Sprunt IV. At the New-York Historical Society: Mary C. Black; Richard J. Koke; Martin Leifer; Ellin Mathieu; Edward H. Santrucek. And also Ruth Annan, Miami, Florida; Herbert R. Axelrod, TFH Publications, Neptune City, New Jersey; Frank C. Craighead Sr., Naples, Florida; George E. Dail Jr., Executive Director, Central and South Florida Flood Control District, West Palm Beach; Edward H. Dwight, Director, Munson-Williams-Proctor Institute, Utica, New York; James S. Haeger, Entomological Research Center, Vero Beach, Florida; Larry G. Pardue, New York Botanical Garden, New York, New York; Glen Simmons, Florida City, Florida; Frank N. Young, Professor of Zoology, Indiana University, Bloomington.

Picture Credits

Sources for the pictures in this book are shown below. Credits for the pictures from left to right are separated by commas; from top to bottom they are separated by dashes.

Cover–Patricia Caulfield. Front end papers 2, 3–Russell Munson. Front end paper 4, page 1–Patricia Caulfield. 2, 3–Patricia Caulfield. 4, 5–Dr. M. P. Kahl. 6, 7–Ed Cooper. 8, 9–Patricia Caulfield. 10, 11–William J. Bolte. 12, 13–James A. Kern. 18, 19–Map by R. R. Donnelley Cartographic Services. 22–Map by R. R. Donnelley Cartographic Services. 24–Karl Knaack from TFH Publications. 29–Patricia Caulfield. 32 through 40–Maps by Walter Johnson. 33 through 41–Russell Munson. 46–Patricia Caulfield. 51 –David Molchos. 54, 55–James H. Carmichael Jr. 63, 64, 65–Dan J. McCoy. 69–James A. Kern. 75–Fred Ward from Black Star. 76, 77–James A. Kern except top right Fred Ward from Black Star. 78, 79–Lynn Pelham from Kay Reese & Associates, Edward Slater. 83–James H. Carmichael Jr. 86, 87–Frederick A. Folger. 90, 91–Patricia Caulfield. 94–Nina Leen. 99 through 107–Paulus Leeser courtesy New-York Historical Society. 113–Jeff Simon. 117–Left Robert Lerner–Treat Davidson from National Audubon Society, right James A. Kern except bottom Patricia Caulfield. 120 through 131–Robert Walch. 137–Dan J. McCoy. 143 through 155–Dan J. McCoy. 160–McFadden Air Photos. 166, 167 –Dan J. McCoy. 171–John Zoiner. 172, 173–Dan J. McCoy. 174, 175, 176–John Zoiner. 177–Dan J. McCoy. 178, 179 –John Zoiner.

Index

Numerals in italics indicate a photograph or drawing of the subject mentioned.

Printed in U.S.A.